EATEN
BY

THE
INTERNET

Eaten by the Internet

Edited by: Corinne Cath
Publisher: Meatspace Press (2023)
Place of publication: Manchester, United Kingdom
Weblink: meatspacepress.com
Design: Carlos Romo-Melgar and John Philip Sage
Commissioning editors: Mark Graham and Joe Shaw
Managing editor: David Sutcliffe
Copy editor: Suzanne van Geuns
Format: Paperback and pdf
Length: 128 pages
Language: English
Product code: MSP231001
ISBN (paperback): 9781913824044
ISBN (pdf, e-book): 978191382405
DOI: https://doi.org/10.58704/dmnx-1r61
License: Creative Commons BY-NC-SA

Contributors (alphabetically): Mehwish Ansari, Yung Au, Corinne Cath, Joan Donovan, Ksenia Ermoshina, Suzanne van Geuns, Gurshabad Grover, Fieke Jansen, Mallory Knodel, Ashwin Mathew, Maxigas, Francesca Musiani, Niels ten Oever, Britt Paris, Jenna Ruddock, Shivan Kaul Sahib, Michael Veale, and Meredith Whittaker.

The publisher has endeavoured to ensure that any URL for external websites referred to in this book are correct and active at the time of going to press. However, the publisher has no responsibility for the websites and can make no guarantee that these links remain live or that the content is, or will remain, appropriate for all audiences.

EATEN
BY

THE
INTERNET

EDITED BY

CORINNE CATH

| 1 | Introduction: eaten by the internet | 7 – 13 |
| 2 | Corinne Cath | |

| 15 | Building muscle in infrastructure | 15 – 23 |
| 17 | Interview with Meredith Whittaker | |

| 23 | Centralising Control, Expanding Industry Reach | 24 |

| 24 | The sprint to plug in the moon | 27 – 34 |
| 25 | Yung Au | |

| 47 | Trust issues | 35 – 42 |
| 49 | Shivan Kaul Sahib | |

| 59 | Confidentiality washing in online advertising | 43 – 48 |
| 60 | Michael Veale | |

| 70 | Mining 'Silicon Holler' for alternatives | 49 – 55 |
| 71 | Britt Paris | |

| 88 | The problem is growth | 57 – 63 |
| 90 | Fieke Jansen | |

| 105 | Government Capture of Internet Infrastructure | 64 |

| 106 | Pipeline ironies | 67 – 74 |
| 108 | Suzanne van Geuns | |

| 131 | The infrastructure of censorship in Asia | 75 – 81 |
| 133 | Gurshabad Grover | |

150	Encryption as a battleground in Ukraine	83 – 88
151	Ksenia Ermoshina	
152	Francesca Musiani	

160   Encryption regulation, and what to do   89 – 95
about it?
161   Mallory Knodel

173   Consumed by the Computer?   96

174   Standing out   99 – 104
176   Niels ten Oever
177   Maxigas

187   The human in the machine   105 – 110
189   Ashwin Mathew

203   Standard politics   111 – 116
205   Mehwish Ansari

208   Towards a public interest internet   117 – 122
210   Jenna Ruddock
211   Joan Donovan

224   Acknowledgements   125

226   Design notes   127

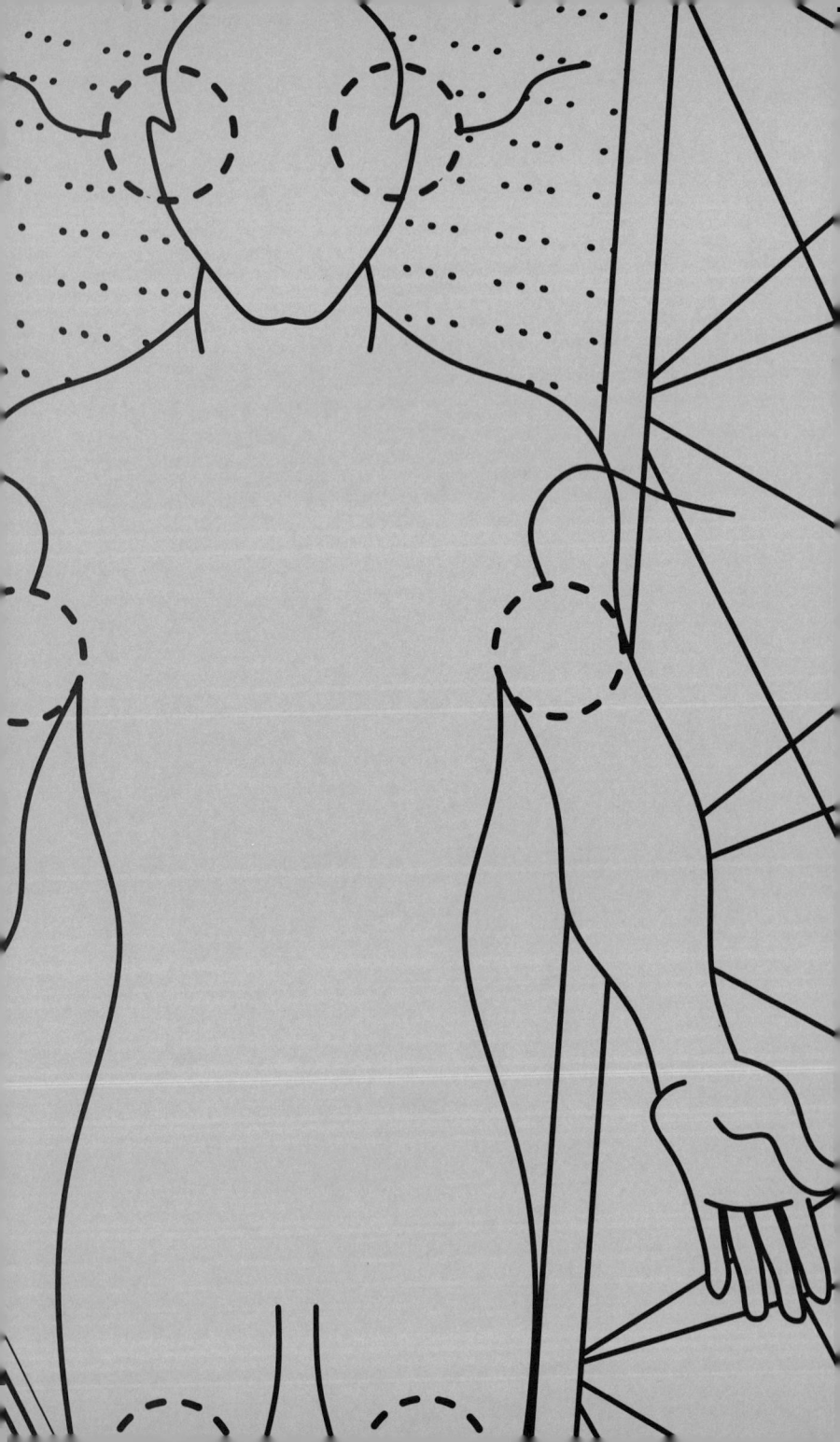

1 Introduction: eaten by the internet

2 Corinne Cath — Corinne Cath is a Post-doctoral Researcher at the University of Delft, fellow at the critical infrastructure lab at the University of Amsterdam, and affiliate at the Minderoo Centre for Technology and Democracy at the University of Cambridge.

3 Keywords: capture, cloud computing, infrastructure, internet governance, politics

4 This is a unique moment: internet technologies are the default infrastructure for society. Not just in the sense of how we communicate with one another, but also in terms of how we organise our social and material environments. But this sizeable domain of technologies rarely attracts attention unless something breaks down. A domain that includes everything from material components such as cell antennas, clouds, chips, data servers, and satellites to less tangible but equally crucial standards and software components, including the operating systems, browsers, and computing power that enable connectivity 5 . And even when this stuff breaks, most internet users do not ask why.

This book begins with the realisation that we ignore internet infrastructure to our own detriment. We cannot mitigate harmful technological developments, if we do not understand how internet infrastructure works or how it is eating the world 6 . Driving these developments is a set of actors and forces whose incentives do not neatly

align with the public interest 🖵. Yet, whoever   7
controls the infrastructure, controls the bounds of
public speech, as well as economic production,
social cohesion, and politics, making infrastructure
a core political terrain in the networked age. To
inquire after the politics of internet infrastructure,
is to inquire after the usually invisible powers that
make our societies what they are.

Internet infrastructure must be made visible as a
force of political power, as it is transforming the
social world, from the bottom up.

If the internet is society's backbone, we need to
think carefully about how power operates through
and around its infrastructures and who wields   8
it 🖵. This book brings together practitioners,
researchers, and techies who help us see how
internet infrastructure is gobbling up society.
These contributors mark two crucial areas of
interest in this inquiry. First, they show that this
power is first and foremost corporate power.   9
Internet infrastructure is increasingly consolidated   10
in the hands of a few tech giants, whose growing
infrastructural reach changes the functioning of
the internet 🖵 and rapidly expands their influence
beyond the realm of communication 🖵. Second,
the contributors reveal how this corporate capture
is interlaced with, and supported by, state priorities
that change public infrastructures to the (further)
exclusion of public values. The 15 chapters
orient the reader away from superficial problems,
toward these two distinct political terrains where
the societal impact of internet infrastructure can
already be seen. The contributors also show what
we stand to gain from thinking about internet

infrastructure and help us imagine a future internet geared toward progressive politics.

CENTRALISING CONTROL, EXPANDING INDUSTRY REACH

The first area of interest involves corporate power over internet infrastructure. Both knowledge and resources, like cloud computing and data analytics capacity, are dangerously consolidated in the hands of a limited number of tech firms, as Signal President Meredith Whittaker clarifies in her interview. Shivan Kaul Sahib demonstrates how corporate capture of internet infrastructure plays out in the domain of online privacy services, where only large existing players can offer key privacy services and only large-scale companies, like Netflix or Google, can afford to buy them—turning privacy into a luxury good. Michael Veale identifies a similar centralising dynamic. In a sharp critique of Google and Apple's hold over the infrastructure and standards required to serve online ads, he notes that many privacy solutions are designed by these companies 'to the detriment of their competitors' rather than to the benefit of their users. This is infrastructural power: being essential, unavoidable, and seemingly too large to topple.

11,12   Internet infrastructure companies are using this power to expand outwards, to grow beyond their original realm of online communication. Various commercial actors are moving aggressively into public functions, like healthcare, education, transport, and rural development ⊡. They ingest these sectors from the inside out, through their infrastructural offerings, with a corporate and colonial logic of extraction, efficiency, and undirected growth. In their chapter, Yung Au analyses patents for orbital data centres and satellites to argue that the corporate ownership of internet infrastructure in space adheres to historical colonial logics. On earth, too, conquest unfolds through infrastructure—but so does resistance, as Britt Paris demonstrates in her chapter on pushback against the extraction of public goods, offered by local internet cooperatives, for private profits of blockchain companies resettling in rural America. All the while, Fieke Jansen argues, the environmental impact of this insatiable expansion remains an afterthought.

GOVERNMENT CAPTURE OF INTERNET INFRASTRUCTURE

This dynamic of corporate expansion and profiting from taking on a political role as a key digital infrastructure provider, is facilitated by state governments. The second area of interest contributors identify therefore concerns the entanglement between internet

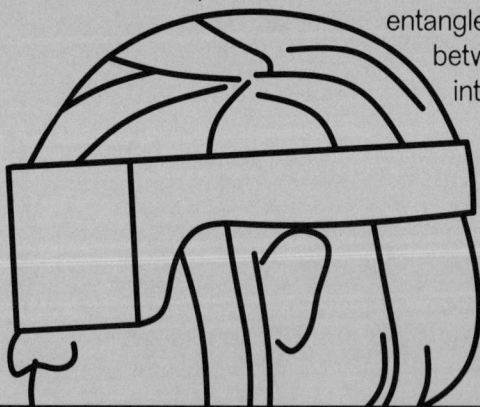

infrastructure and state surveillance projects. The chapters outline the pernicious ways in which governments capitalise on internet infrastructure to further their political agendas. States can do this by exercising control over the corporations, such as Internet Service Providers (ISPs) and telecommunication companies, that own and operate the internet's material infrastructure. When

13 ⟿⟶ governments make an 'infrastructural turn,' 💬. they make these companies responsible for censoring and filtering information, which complicates public oversight and accountability.

Power works through infrastructures old and new. Highlighting continuity, Suzanne van Geuns connects the present-day 'pipelines' of internet infrastructure—and the forums that use them to push online hate—to the much older infrastructure of American imperialism in Asia. In their chapter, Gurshabad Grover argues that more knowledge is needed on the current and new infrastructural implementation of government control in Asia, to resist it. Infrastructure also clearly plays a key role in contemporary conflict. In their contribution, Ksenia Ermoshina and Francesca Musiani show how Ukrainians are adapting their digital security practices in response to the Russian invasion. Infrastructure comes into view as both a pathway to (colonial) domination and a site of defiance and possibility.

When states and companies play into one another's hands, the public suffers. In her chapter on the myriad of legislative attacks on end-to-end encryption across the globe, Mallory Knodel offers powerful examples of the constriction of people's political rights at the behest of governments. This book uses the troubling enmeshment of the state and private sector in internet infrastructure to highlight the need for bottom-up interventions.

CONSUMED BY THE COMPUTER?

What might effective interventions look like? To mobilise a networked public empowered to foster infrastructures that serve their interest, we must get concrete about the harms posed by current infrastructural developments—and be just as concrete about alternatives. The chapter contributed by Niels ten Oever and Maxigas of the critical infrastructure lab provides one method for democratic control from the ground up: 'walking the city' and immersing yourself in internet infrastructure to notice how it undergirds every aspect of social life. The chapters of Mehwish Ansari, Jenna Ruddock and Joan Donovan, and Ashwin Mathew, lay out how the networked governance of technology, with civic organisations, social movements, and informal networks of internet governance professionals collaborating around the prevention of tech harms, can help us mount resistance to the corporate-governmental power nexus.

Rooting contemporary technology debates in infrastructure—that is, in key aspects of its invisible functioning where power congeals—is a crucial refocusing of our political energies. Only when we understand the role that it plays can we begin to ask what might serve us better; how can we ensure our infrastructures sustain us, rather than consume us?

---

14   Notes:

5.   Gürses, S. 2020. Programmable Infrastructures - Seminar on Programmable Infrastructures. TU Delft. https://www.tudelft.nl/evenementen/2020/tbm/programmable-infrastructures

6.   This title nods to the 2011 Wall Street article, in which Marc Andreesen proclaimed that software was eating the world https://www.wsj.com/articles/SB10001424053111903480904576512250915629460 The internet plays a crucial role in the gluttonous expansion of software and established this network-of-network as the infrastructural backbone to all matters of societal institutions and processes, like commerce, politics, education, and health.

7.   https://www.brookings.edu/techstream/how-hate-speech-reveals-the-invisible-politics-of-internet-infrastructure

8.   https://corinnecath.com/wp-content/uploads/2021/09/CathCorinne-Thesis-DphilInformationCommunicationSocialSciences.pdf

9.   https://blog.apnic.net/2020/12/04/unpacking-a-flattened-internet/

10.   Recent academic research suggests that the dominance of a limited number of tech companies is leading to a bifurcation, whereby these companies no longer rely on the public internet to route their traffic but increasingly move it across private networks, raising concerns for the future of the public internet—a trend various chapters in this book touch on. See also https://papers.ssrn.com/abstract=3910108

11.   https://research.tilburguniversity.edu/en/publications/digital-disruption-or-crisis-capitalism-technology-power-and-the-

12.   https://www.ihub.ru.nl/project/spheretransgressionwatch.page

13.   For further background on 'the turn to infrastructure' see: Musiani, F., Cogburn, D.L., DeNardis, L. and Levinson, N.S. 2015. *The Turn to Infrastructure in Internet Governance*. New York, NY: Palgrave Macmillan US.

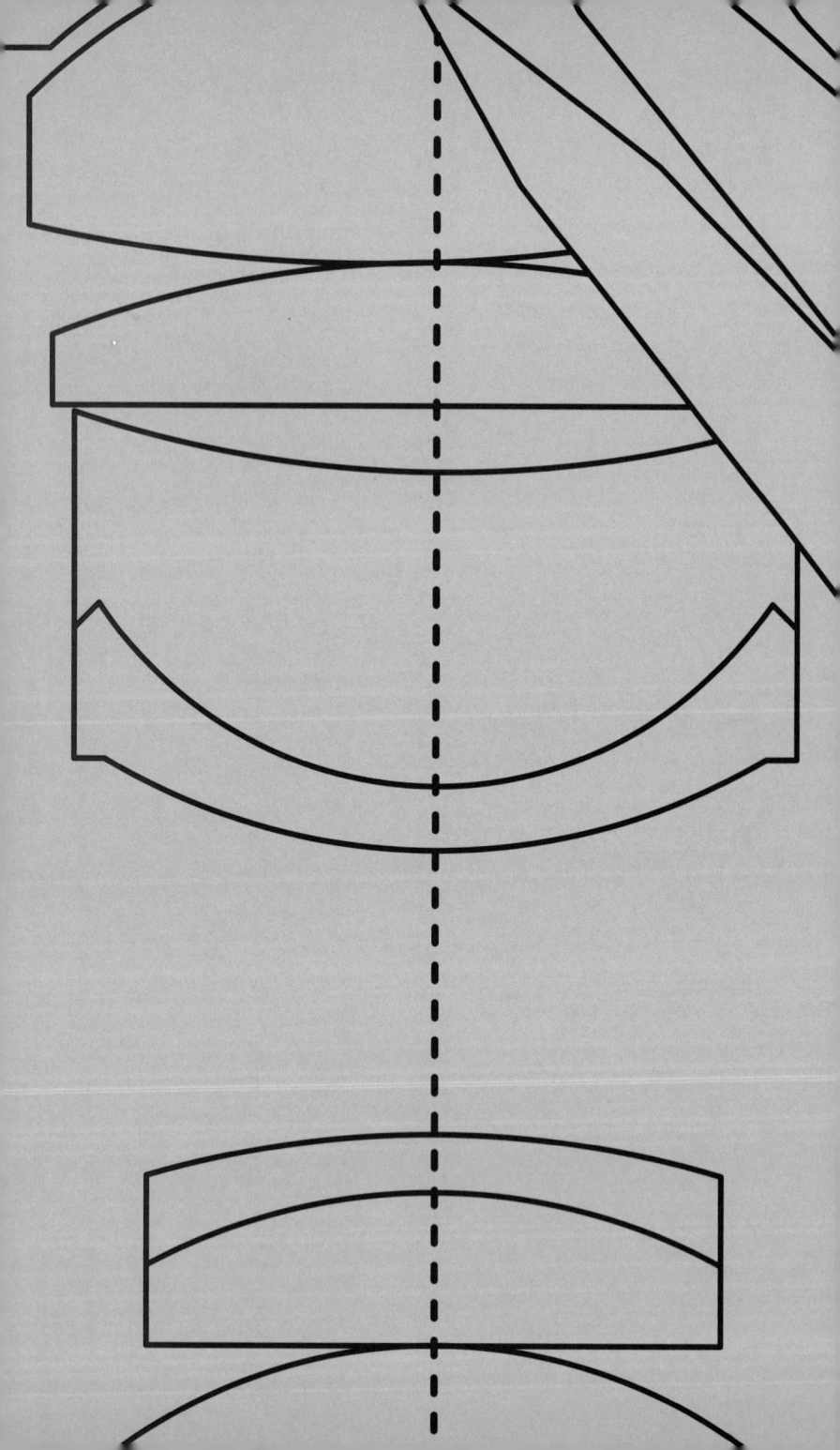

Building muscle in infrastructure

Meredith Whittaker on Signal's space in tech

Interview with Meredith Whittaker

Meredith Whittaker is the president of the Signal Foundation and serves on their board of directors. She was the co-founder and faculty director of the AI Now Institute. She also served as a senior advisor on AI at the Federal Trade Commission. Whittaker was employed at Google for 13 years.

18 Keywords: big tech, capital, open source, Signal, surveillance

---

19 The internet is overwhelmingly geared toward surveillance for profit. This business model requires ever more sophisticated and large-scale infrastructures to operate, and it is possession over these infrastructures, entrenched market reach, and large stores of data that can be continually refreshed via such market reach, which consolidates power in the tech sector. One place where alternative infrastructures exist in some

20 form, however, is the encrypted messaging app Signal ○. Meredith Whittaker, Signal's President, spoke with us about power and the tech sector, the ways in which the open tech space undermines itself, the need to build strength in the movement, and the tactics Signal uses to sustain a private and safe alternative in a tech sector organised around the surveillance business model.

I will never forget the moment at a 2015 conference when you told me that the

infrastructure group was one of the most powerful groups at Google, but that they were kept out of the public eye. Can you tell me a bit more about the politics of hiding—why do companies keep under wraps how much they spend on data centres and pipes?

In general, the people who have real power don't have to be public. That is a privilege. Let me offer a somewhat smug academic answer, drawing on Bowker and Star ⭕ one of the definitional components of infrastructure is its invisibility. It only becomes visible in ruptures: when it breaks, or there is an issue, or a sudden mismatch with the context.

21

One of the key mythologies of tech is that networked computation, and the explosion in consumer-facing tech, is the result of scientific innovation. In this story, there are two boys in a garage with an idea. And by the transitive property of 'science' and 'innovation' and the brilliance emanating from these boys as well as the rarefied geography of Silicon Valley, this spark of a brilliant idea becomes a service we all use. The economic model is hidden. The products are 'free' because these intelligent innovators finally found a formula to provide needed services while profiting ethically. They 'innovated' technology and the formula for ethical billionaires! This sounds very cynical, and I *am* a bit cynical about it, but back in the mid-2000s people did talk like this, with a straight face.

The free products, and the breathless narrative that conflates the products developed and deployed by tech companies with 'innovation' and 'scientific progress', distracts the public from the fact that offering these products requires massive resources and infrastructures that are unavailable to most. This is what companies hide: is it innovation or is it capital? In hindsight, it becomes clear that the people and companies that now dominate the tech industry benefited more from first-mover advantage than from technical brilliance.

These companies, like Google, keep infrastructure out of sight because infrastructure is capital, so to speak. And if it is capital and not innovation that propels their business, then perhaps it should be regulated, perhaps it does exist within a more familiar political economic framework that doesn't require genius and computer science expertise to understand and grapple with. Then perhaps what we are looking at is an uneven playing field—in which Google's and other Big Tech companies' advantage is built on their concentration of capital and resources. And the fantasy of an even playing field, the meritocratic myth, is important beyond just providing justifications for the fact that only certain types of people get promoted, etc. The myth of meritocracy helps conflate technical products with scientific and moral progress— what you see is the best, the smartest, the most capable. When actually, the advantage is largely infrastructural—it is not about being better at science or ideals or tests or whatever. Hidden in the background is the reality of a self-reinforcing monopoly: the more infrastructure you have, the more data you have, and the more people using your devices and platforms, all these things compound into a monopoly that these companies do significant work to naturalise, via narratives of meritocracy and scientific innovation.

Infrastructure is definitionally hidden, and it is also where the money is. If you look at the money, you see that the story of tech is less Athena popping from Zeus's head with a brilliant idea and more the product of concentrated power.

You have always said that you have little affinity for the industry, but that you're interested in power. What do you mean by that?

Well, my knowledge of tech, and my education, took a very unorthodox path. I was educated 'in tech' while working at Google, but I was always reading on my own. I read and at the same time, my community of practice was the people building things, taking the status quo of the tech sector as given, extolling the virtues of organising the world's information, making information free, etc. And I read in parallel to try and get a grasp on what made me uncomfortable about the common sense in tech. And that helped direct my attention to the material, to the infrastructural layer where things got real and concrete, and the promises of magic were executed (or not) at the layer of labour and raw materials and always-almost-broken operating systems. We cannot talk about internet infrastructure outside of capitalism, and we cannot talk about capitalism right now without cognising internet infrastructure.

But what is internet infrastructure? Of course, there are data centres and massive server farms. There are devices produced by a handful of companies, and operating systems, which are essentially controlled by two companies: Google and Apple—and then there is Linux (also largely funded by these Big Tech companies). But we must also include labour and practice, content moderation, device manufacture in Shenzhen, rare earth mineral mining, etc.—all of which is infrastructure.

I am interested in that complicated assemblage because I don't think it is possible to understand how things work, or how power flows, without understanding this infrastructural layer. What affordances sit below the thing we are seeing on the surface? Who owns those affordances? These are the questions we should be answering.

You have always been at the cutting edge of new developments: the development of Measurement-Lab at Google, the tech workers' union, the subsequent founding of AI Now, and currently as the President of Signal. Can we take your move to this organisation as a sign of what lies ahead for the tech sector?

Signal needs to exist and I'm so proud to be able to work to ensure it does. Take what is happening with Twitter. The scenario in which a badly intentioned actor suddenly gains access to a tech company and all its data used to be the hypothetical that people like me brought up in public talks about privacy, to try and get people to care. Now it's real.

The surveillance business model that characterises the tech sector trends towards natural monopoly, as we've discussed briefly. We now have a handful of companies that have entrenched their power. Their infrastructures and capabilities thread through our institutions and lives in ways that we have no choice over, collecting and creating data that has significant power over us. So there needs to be a way to communicate safely, privately, and intimately with each other. A world where we are persistently surveilled is a world where power asymmetries are calcified. This is not a new insight, but we are in an unprecedented and scary time and so I see this job as the President of Signal as my service in a larger struggle.

The other side of that is that academia is collapsing. It was collapsing last year; it is collapsing now. Look at Florida. Look at the College Board. In the US, we are seeing whole humanities programmes gutted. Everything radical that does more than moulder in a barely read journal gets pushed out because it scares donors, and universities privilege donors above all. So where does critical work live? That is a really big question, especially in engineering schools, where you cannot do cutting-edge engineering work without massive grants and infrastructural access from tech companies.

We are seriously outmatched. And the danger, especially in the 'critical open tech' funding space, is that there is an incentive to claim that you did more than you did because funders like impact and the sector is reliant on shoestring grants. That is dangerous because it creates the illusion that what we are doing is sufficient. It's not true. We are not going to survive unless we actually build some muscle.

Where does the critical work happen and how do you think we can build and flex more muscle?

We build and flex that muscle by being serious about what we are doing. It is not serious to invite

Eric Schmidt into a room with critical tech funders and give him the map to charter the future of open tech. We are not actually doing critical work unless we recognise that we are up against these people. This is a battle of power. The status quo is not a misunderstanding that we can paper over with a 'new idea'. Ideas are cheap and we have had them for years; what we don't have is power and sufficient resources.

In order to get power, we need to organise. The critical tech space, the civil society organisations—it is all highly competitive. This competition is a dynamic that is fuelled by academia and funders, and it mirrors the tech industry. We are not sharing the map so we can figure out how we can push back. No, we are hoarding the map so we can be the first to coin a silly term that everyone uses. Not because we are bad people, but because it is what we are incentivised to do, and ultimately, we need health insurance and to pay rent.

What does it take to sustain open-source alternatives like Signal and build different political-economic models, like Signal is doing?

It takes money. You need to pay your staff, your hosting, your registration, and your computational processes—which are significantly more expensive when they are privacy-preserving. Dumping everyone's information in a relational database is much cheaper. The core problem is how to pay for operations when you are not monetising surveillance in one way or another.

Signal is lucky because Brian Acton has been generous enough to provide a significant runway for Signal, with a $50m loan. But this doesn't solve the problem. It costs tens of millions of dollars a year to run Signal, and that's lean. One thing that Signal does have, fortunately, is the network effect. It is tried and true: millions and millions of people across the globe rely on it, so we have a way to reach people and ask them to kick in. We are now starting to experiment with asking people directly to donate to Signal. That is not a model I would prescribe for an entire industry, but it is a model that we think will work for us.

We are also structured as a non-profit, which means that we aren't incentivised to sell out the people who rely on us. Well, we could, but no one on the executive team would benefit, because we would have to put the money in another charitable venture. And we're not being pushed by a board and investors to prioritise profit—something that in an industry built on the surveillance business model is generally antithetical to privacy. We also don't have data that might incentivise people to acquire us. These are structural ways in which we have tried to protect ourselves and our integrity.

What role do you foresee for Signal in the political and economic reality you just sketched out? You are one of the few organisations that is not trying to profit in tech. What does it take to sustain open-source alternatives like Signal and build different political-economic models, like Signal is doing?

Millions of people love it and use Signal. We are at a moment where I am optimistic about people's willingness to support something like Signal. The mask is off the tech industry. No one is confused about how they make money, and people care about privacy. This is also a time of multiple legislative attacks on privacy. These attacks are more sophisticated than they have been in the past, wrapped in the guise of preventing child exploitation or in the 9/11 trope of terrorism. Signal provides people with a choice they do not have anywhere else and soon might not have anywhere else.

---

22   Notes:

20. https://signal.org
21. Bowker, G. C. and S. L. Star. 2000.
     Sorting Things Out: Classification and Its Consequences. Revised edition. Cambridge, MA: The MIT Press.

PART 1

CENTRALISING CONTROL,

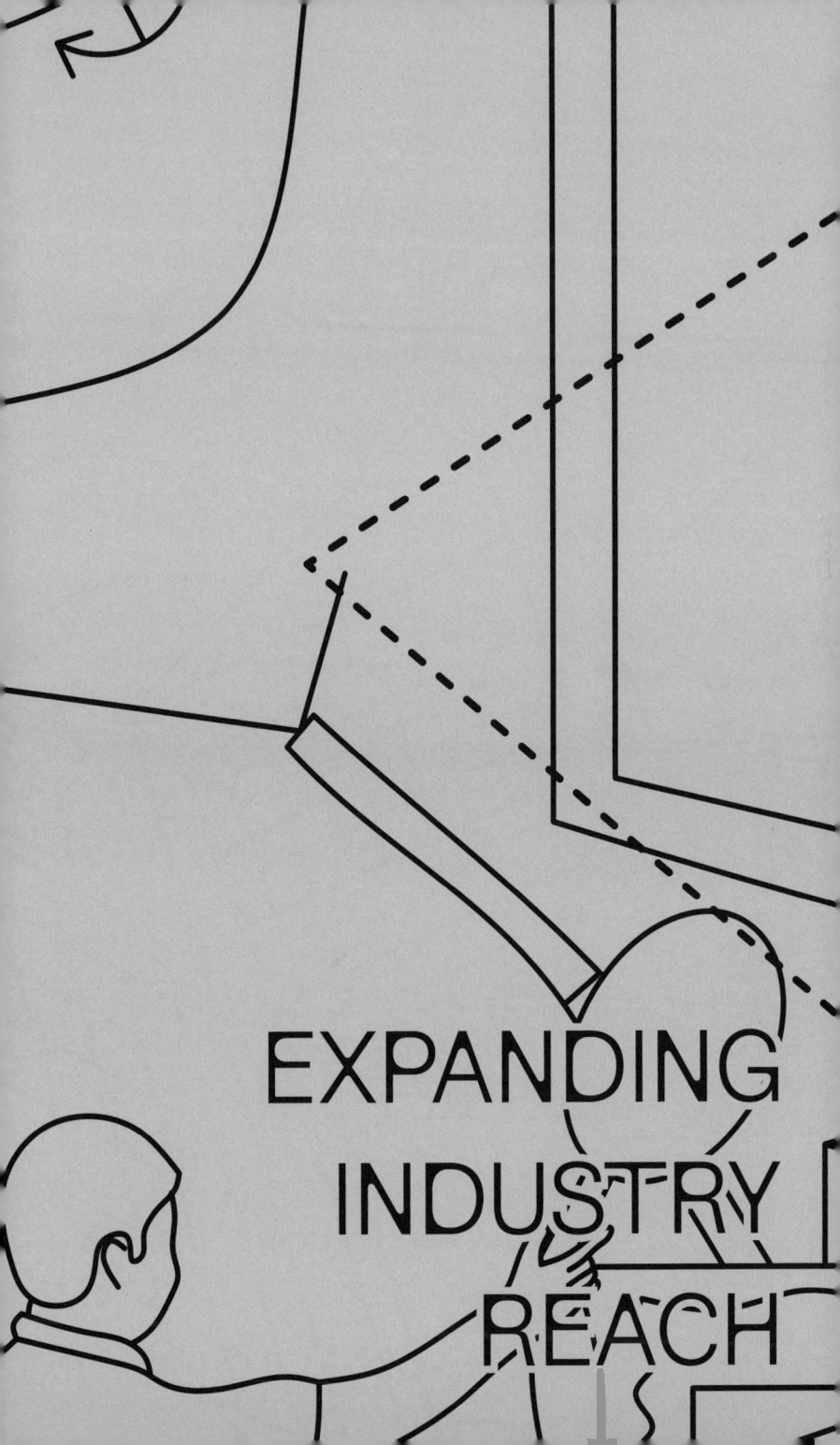

EXPANDING INDUSTRY REACH

24 The sprint to plug in the moon

25 ^ Yung Au — Yung Au is a DPhil researcher at the University of Oxford, and an Associate Lecturer in University Arts London.

26 Keywords: Amazon, pollution, post-colonialism, satellites, space internet

27 *'The network architecture described herein can be used to provide global high-speed internet service from above the sky through various aircrafts such as high-altitude airships, satellites, space stations, and eventually the lunar data centres.'*

2018 Amazon patent for 'satellite-based Content Delivery Network in an extra-terrestrial environment'

Infrastructures for storing and processing data are expanding in outer space. At present, there are more than 7,500 active satellites, with additional inactive satellites still in orbit. Satellites can be multifunctional, including acting as small-scale data centres when fitted with a central processing unit. As such, many mini data centres are currently in orbit. The largest modular space station, the 420-tonne International Space Station (ISS), will circle the Earth until its retirement in 2024. The ISS itself holds about 100 laptops and acts as a data centre in various ways. The explosive infrastructural growth of recent years involves 28 claiming and interrupting environments and social systems for data needs ♀.

When these developments seep into cosmic space, the communities implicated are all of us. It is not just governments with ambitious galactic plans. For instance, in 2019, Amazon patented a space-based Content Delivery Network ♀. The patent outlines the plans for a geographically distributed network of data centres in outer space and, eventually, lunar data centres as part of Amazon Web Services. As companies, governments and coalitions continue their expansionist dreams of internet infrastructures in interstellar environments, what politics will unfold? ♀ What will the race to fill the crowded skies reveal about our priorities, dreams, and assumptions of progress?

COLONIAL CLAIMS TO THE COSMOS

The rush to claim the cosmos brings with it colonial logics ♀. Colonialism is an ongoing system of power predicated on extractive domination over land, resources, and people ♀. While many actors are contending with their colonial pasts, colonial ways of thinking persist —as do the material inequalities this system sanctions. Coloniality endures today which can be understood as 'patterns of power that emerged as a result of colonialism, but that define culture, labour, intersubjective relations, and knowledge production well beyond the strict limits of colonial administrations' ♀. Two ways in which coloniality persists in outer space are first, in the claiming of outer space, and second, in the afterlife and environmental impact of these cosmic infrastructures.

29

30

31

32

33

LUNAR REAL ESTATE

34 → Coloniality persists in thinking about space through what Deondre Smiles calls fantasies of *terra nullius* or no man's land ♀. The cosmos becomes a frontier like the land-based frontiers colonial settlers sought out; spaces to be claimed for extraction, further settlement, and the preservation of the status quo on Earth. Infrastructures for claiming prime real estate and resources extend to

35 → orbital pathways, acres of the moons and planets, and more ♀.

Furthermore, many speculative projects have already claimed the moon. Aside from Amazon, French company Thales, in collaboration with the Italian and American Space Agencies, is

36 → examining the possibility of lunar human quarters and computing infrastructures ♀. Lonestar Data Holdings, a US start-up, has also announced plans to build data centres on the moon—or

37 → what they call, 'the world's ultimate offsite backup location' ♀. Another US-based start-up called ConnectX has crafted blueprints to store crypto-currencies off-planet, announcing that 'space gives ConnectX an unfair advantage. Economically,

38 → we do not have to pay for real estate, electricity, cooling, staff, or security' ♀.

These examples showcase overarching frameworks of space-based infrastructural projects. Narratives include presenting extra-terrestrial bodies as outside of Earth-based capitalist and ownership systems where these vast, empty spaces are up for grabs for anybody with such ambitions. Similarly, is the idea that outer space contains lucrative resources that remain untapped

and
untaken. As
Smiles writes, colonial
logics are most overt in the
invocation of outer space as a 'wild
west', a 'frontier to be tamed'. The question,
under this framing, becomes: if we do not exploit
this emptiness—who will?

COLONIALITY OF SPACE POLLUTION

Coloniality persists in the afterlives of space
infrastructure projects whereas Max Liboiron
argues, in many cases, pollution is colonialism ♟.   39
Like all data infrastructures, space endeavours
incur uneven maintenance and environmental
costs. Terrestrial projects are increasingly
contending with the environmental costs of
computing, from cryptocurrencies to large
language models ♟. Pollution processes here   40
often fit what Liboiron calls the 'permission-to-
pollute' framework. Born of colonial land relations,
this system assumes an entitlement to land and
atmospheres for settler-colonial goals that benefit
some but harm others. Under this logic, the task
is to manage inevitable pollution rather than to
eliminate it—or even to question whether this is a
just system in the first place. Such logics continue
in outer space.

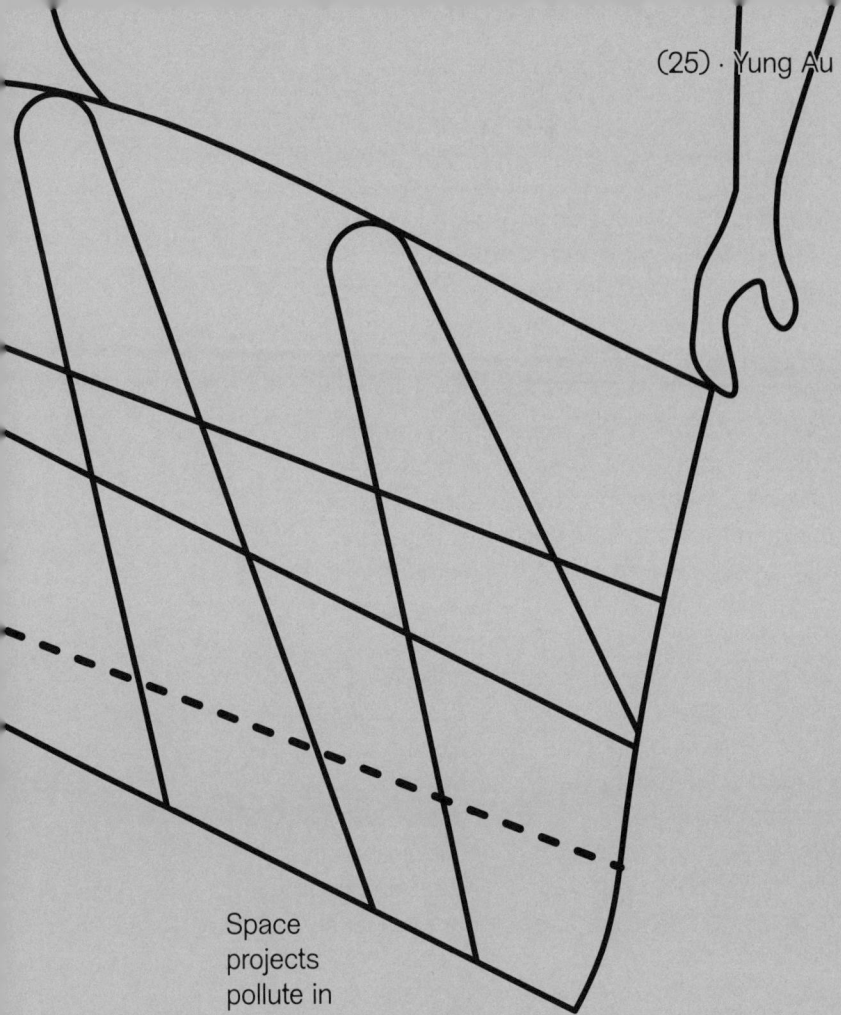

Space
projects
pollute in
many ways. Satellite
constellations create light
and sound pollution, profoundly
affecting cultural, scientific, and
41   social practices that rely on dark skies
and silence ♁. Decommissioned satellites left
42   in orbit become space junk and create a risk of
collision events ♁. Decommissioned satellites
can also be designed for demise, which means
that they burn up upon re-entering Earth where
environmental consequences are complex and
include the potential for accidental geoengineering.

Space debris can alternatively be removed through controlled re-entry, where they are most commonly buried in the South Pacific Ocean Uninhabited Area ⚇. While more than 260 satellites have been buried, the practice remains contentious as concerned actors have pointed out the unknown risks to larger ecosystems ⚇.

43

44

Even well-intentioned science can reproduce colonial logics. The *terra nullis* fantasies of outer space tuck away the environmental impacts in the race to fill the vast expanse. The brunt of the cost of space exploration is already unevenly distributed ⚇. Furthermore, these environmental concerns affect global communities, such as the potentially irreversible geoengineering of Earth's atmospheres. However, the decisions regarding these risks are made by a much smaller consensus. How do we decide when a pursuit is worth change at the planetary scale? Is, as the Amazon patent puts it, the pursuit of ubiquitous internet and real time gaming worth changing the face of our only moon forever? As the ISS is laid to rest in its watery grave in 2024, what would it look like to take a more expansive understanding of the afterlives of our starry fantasies?

45

COSMIC FUTURES FOR SALE

Coloniality persists in the expansion of data infrastructures in outer space, from their speculation stages to their many afterlives. This includes the supposed entitlement to extra-terrestrial space, radio-wave spectrums, the ocean, and the night sky by the constellation of actors in this exclusive race to space. Here, colonial logics claim vast celestial spaces that should be considered as resources and natural spheres meant to serve the many rather than turn a profit for the few.

There are many unknowns about future developments beyond Earth, but it is important to extend foresight over these issues. The diverse work from post-colonial and de-colonial activists and scholars provides a robust foundation from which to reassess the cosmic futures we are being collectively signed up for. It is important, at this critical juncture of time, to heed the cautionary tales of colonialism and its myriad harms. Understanding more precisely the ways in which coloniality extends into projects of data infrastructures will help us extricate what interstellar infrastructural injustices might look like—and, perhaps, how to prevent a world where the moon is sold to the highest bidder.

---

46   Notes:

28. Amoore, L. 2018. Cloud geographies: Computing, data, sovereignty. *Progress in Human Geography* 42 (1) 4-24. doi: https://doi.org/10.1177/0309132516662147; Hogan M. 2018. Big data ecologies. *Ephemera - Theory and Politics in Organization* 18 (3) 631-657; Hu, T.H. 2015. *A Prehistory of the Cloud.* MIT press.

29. Amazon. US Patent for Satellite-based Content Delivery Network (CDN) in an extraterrestrial environment (Patent # 10,419,106). https://patents.justia.com/patent/10419106

30. Gorman, A. 2005. The cultural landscape of interplanetary space. *Journal of Social Archaeology* 5 (1) 85-107; Klinger, J.M. 2018. *Rare Earth Frontiers: From Terrestrial Subsoils to Lunar Landscapes.* Cornell University Press. https://library.oapen.org/handle/20.500.12657/30764; Maile, D.U. 2018. Precarious Performances: The Thirty Meter Telescope and Settler State Policing of Kānaka Maoli. *Abolition Journal.* https://abolitionjournal.org/precarious-performances; Cath, C. and Lewis, B. 2021. *Space-Cowboys: What Internet history tells us about the inevitable shortcomings of a tech-bro led Space Race.* Tech Policy Press. https://techpolicy.press/space-cowboys-what-internet-history-tells-us-about-the-inevitable-shortcomings-of-a-tech-bro-led-space-race

31. Smiles, D. 2020. The Settler Logics of (Outer) Space. *Society and Space*. https://www.societyandspace.org/articles/the-settler-logics-of-outer-space; Tuck, E. and Gaztambide-Fernández R.A. 2013. Curriculum, Replacement, and Settler Futurity. *Journal of Curriculum Theorizing* 29 (1). https://journal.jctonline.org/index.php/jct/article/view/411

32. Wolfe, P. 2006. Settler colonialism and the elimination of the native. *Journal of Genocide Research* 8 (4) 387-409. doi: https://doi.org/10.1080/14623520601056240; Veracini, L. 2014. Understanding Colonialism and Settler Colonialism as Distinct Formations. *Interventions* 16 (5) 615-633. doi: https://doi.org/10.1080/1369801X.2013.858983

33. Maldonado-Torres, N. 2007. On the Coloniality of Being: Contributions to the Development of a Concept. *Cultural Studies* 21 (2–3) 240–70. https://doi.org/10.1080/09502380601162548.

34. Smiles, D., at note 4.

35. Au, Y. 2022. Data Centres on the Moon and Other Tales: A Volumetric and Elemental Analysis of the Coloniality of Digital Infrastructures. *Territory, Politics, Governance*, 1–19. https://doi.org/10.1080/21622671.2022.2153160; Purity, N. 2020. Spectrum & Orbital Slotting – A case for African Countries. Space in Africa. https://africanews.space/spectrum-orbital-slotting-a-case-for-african-countries; Rothblatt, M.A. 1982. Satellite Communication and Spectrum Allocation. *The American Journal of International Law* 76 (1) 56-77. doi: https://doi.org/10.2307/2200975

36. Menear, H. 2021. NTT and SKY Perfect are building data centres in space. *Data Centre Magazine*. https://datacentremagazine.com/technology-and-ai/ntt-and-sky-perfect-are-building-data-centres-space; Moss, S. 2021. Thales to study "Lunar Data Center" as part of Artemis program. *Data Center Dynamics*. https://www.datacenterdynamics.com/en/news/thales-to-study-lunar-data-center-as-part-of-artemis-program

37. https://www.lonestarlunar.com/services

38. Donoghue, A. 2018. The Idea of Data Centers in Space Just Got a Little Less Crazy. *Data Center Knowledge*. https://www.datacenterknowledge.com/edge-computing/idea-data-centers-space-just-got-little-less-crazy

39. Liboiron, M. 2021. *Pollution Is Colonialism*. Duke University Press. doi: https://doi.org/10.1515/9781478021445

40. Hogan, M. 2018. Big Data Ecologies. *Ephemera – Theory and Politics in Organization* 18 (3) 631–57; Akese, G.A. 2019. Electronic Waste (e-Waste) Science and Advocacy at Agbogbloshie: The Making and Effects of The World's Largest e-Waste Dump. PhD Thesis, Memorial University of Newfoundland, 2019; Bender, E.M. et al. 2021. On the Dangers of Stochastic Parrots: Can Language Models Be Too Big? In Proceedings of the 2021 ACM Conference on Fairness, Accountability, and Transparency, FAccT '21, New York: NY, pp. 610–23, https://doi.org/10.1145/3442188.3445922

41. Chakrabarti, S. 2021. How many satellites are orbiting Earth? Space.com. https://www.space.com/how-many-satellites-are-orbiting-earth; Venkatesan, A., Lowenthal, J., Prem, P. and Vidaurri, M. 2020. The impact of satellite constellations on space as an ancestral global commons. *Nat Astron*. 4 (11) 1043-1048. doi: https://doi.org/10.1038/s41550-020-01238-3

42. Boley, A.C. and Byers, M. 2021. Satellite mega-constellations create risks in Low Earth Orbit, the atmosphere and on Earth. *Sci Rep*. 11 (1) 10642. doi: https://doi.org/10.1038/s41598-021-89909-7

43. See http://www.nasa.gov/feature/faq-the-international-space-station-2022-transition-plan

44. De Lucia, V. and Iavicoli V. 2018. From Outer Space to Ocean Depths: The Spacecraft Cemetery and the Protection of the Marine Environment in Areas beyond National Jurisdiction. *Cal W Int'l LJ*. 49 (2) 345-390.

45. Uahikeaikalei'ohu Maile, D. 2018. Precarious Performances: The Thirty Meter Telescope and Settler State Policing of Kānaka Maoli. *Abolition Journal*. https://abolitionjournal.org/precarious-performances; Young, M. 2013. 17-Day ALMA Strike Ends in Resolution. Sky & Telescope (blog), https://skyandtelescope.org/astronomy-news/17-day-alma-strike-ends-in-resolution

47 Trust issues

48 Internet privacy and centralized infrastructure

49 Shivan Kaul Sahib   Shivan Kaul Sahib is a privacy
engineer and researcher active in
internet privacy standardisation at
IETF and W3C. He can be found
online at https://shivankaul.com

50 Keywords: Content Delivery Networks (CDN), Google,
privacy, protocols, trust

51   In the internet ecosystem, Content Delivery
Networks (CDNs) are the biggest piece of
infrastructure you haven't heard of. CDNs are
systems of servers spread out across the world
that cache, or store, content (like images, videos,
and web pages) and serve it to users from their
nearest location. This reduces load times and
improves website performance. For example, if
you live in New Delhi, when you do a search on
Google, the website that is sent to your browser
is actually served from a server in India, not from
Google's HQ in Mountain View, California. CDNs
sell scale and pool content; by their very nature,
they centralise the internet.

Deploying thousands of servers across the
world is expensive, which means there are only
a handful of mainstream CDNs. But CDNs do
not just wield infrastructural power in deciding
what users can see and access online; they
are also increasingly prominent in the field of
internet privacy, where they are starting to serve
the role of trusted infrastructure in new privacy-

preserving protocols. While
these protocols are important
for privacy, the reliance on
expensive infrastructure has
the effect of making them
deployable only by large
tech companies. This has
ramifications for the politics   of access
to privacy on the internet and a danger that smaller
organisations acting in the public interest will not
be able to afford to provide privacy for their users.

NEW PRIVACY PROTOCOLS

The central privacy problem with traditional
encrypted connections between your phone and

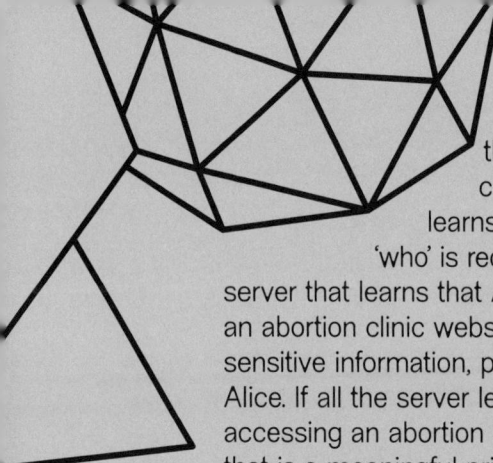

the server that stores the
content is that this server
learns both 'who' and 'what':
'who' is requesting 'what' content. A
server that learns that Alice ('who') has visited
an abortion clinic website ('what') can leak that
sensitive information, possibly endangering
Alice. If all the server learns is that *someone* is
accessing an abortion clinic website, however,
that is a meaningful privacy improvement. This
improvement is even more pronounced if a *lot* of
people are accessing that website. The anonymity
of crowds is good for user privacy.

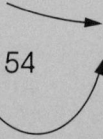

Breaking this link between 'who' and 'what' in
users' internet access is a major principle in recent
privacy protocols and products. Oblivious DoH,
Oblivious HTTP, and Apple's Private Relay
all build privacy protections into the technical
infrastructures that connect networks, users, and
servers. They do this by adding one or more
servers called 'relays' between the device that
requests the content and the server that stores it.
This new relay server doesn't learn 'what' you're
accessing, but it *does* know 'who' you are. It *relays*
your request on your behalf. The final server learns
'what' you're accessing—it has to because it has to
send the content back to you—but it doesn't learn
'who' you are because it only talks to the relay.
No one party learns everything, which makes it
impossible to say definitively that a particular user
did something.

The underlying principle here is to divide up
the trust. Other solutions that draw on this
principle propose a form of privacy-preserving
measurement, in which sensitive user data is split

52

53, 54

up and sent to two separate servers. That way, no one server ever learns one user's specific data, but it is still possible to get aggregate statistics from everyone who contributed. Approached as a matter of trust, this multiple-server approach to privacy divides trust between two parties. User privacy is protected *as long as the two parties don't collude,* i.e. by sharing information, because to do so would be to defeat the point of the separation. This approach creates the need for a *trusted third party:* an organisation or company that agrees to serve as one of the parties and promises not to collude with the other party.

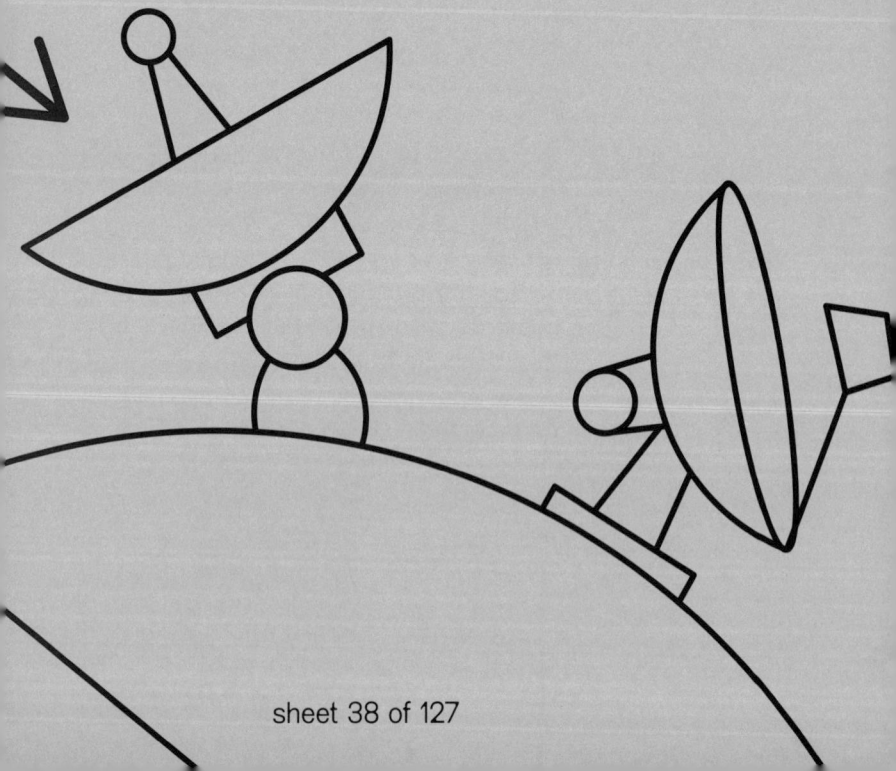

TRUST-AS-A-SERVICE

To be a trusted third party is to take on an infrastructural role. The business model is that those who want to provide users with content while preserving their privacy will pay for the service. In other words, trusted third parties sell trust-as-a-service. Trusted third parties are not a new concept, but what distinguishes the role of a trusted third party in new internet privacy protocols is the traffic component. If there is protection in the crowd, as these protocols assume, the trusted organisation needs to be able to handle a lot of users. This makes Content Delivery Networks, or CDNs, appear as a natural fit to take on this role.

CDNs already provide infrastructural services at a large scale to companies on the internet. They are also expensive because they provide a service that only big players require—for example, a content provider like Netflix needing to distribute its content worldwide to meet global demand. The need for a CDN only arises when such demand is high, and revenue is flowing; it is hardly surprising that CDNs therefore take a big slice of the profit in exchange for expanding the pie.

$$$

Protecting users' privacy, however, should not be expensive. It should be cheap and easy. While it is great that privacy-enhancing protocols are being developed, with increasingly innovative ways of splitting up the trust, we should consider how expensive or difficult it is to implement and run them. If the infrastructure bill is so high that only a company the size of Google could deploy or implement these privacy standards, that is a serious problem.

It puts users in a position where their only choice for retaining their privacy on the internet will be to trust Big Tech *and* a commercial third party. This trusted third party is completely invisible to the users who are supposed to trust them. Even worse, there are very few infrastructure companies in the world capable of running such a service—further centralising internet infrastructure.

Should users just embrace CDNs as their only chance at workable privacy protocols online? No; the good news is that we have a precedent for relying on organisations that act in the public interest in these situations. The ubiquitous adoption of an older privacy-enhancing protocol, the HTTPS protocol, which ensures that your connection to a website is secure, was largely enabled by the Let's Encrypt project, which effectively acted as a trusted third party and issued the digital certificates this protocol required for free. Crucially, Let's Encrypt is run by a non-profit using donations and grants; it effectively acts as public interest infrastructure.

55

While it remains to be seen how the infrastructure-reliant privacy protocols currently in development end up being deployed and with what business relationships, the active participation of ISRG (the parent organisation of Let's Encrypt) in developing these protocols is heartening —one hopes that they would take on a similar non-profit-driven role in the actual operation of these systems. There is also active work being done on developing cryptographic privacy protocols that are cheap and easy to run

56

57

while guaranteeing similar privacy properties. This should be encouraged and supported. We can expect to see these protocols starting to be widely deployed in the coming few years.

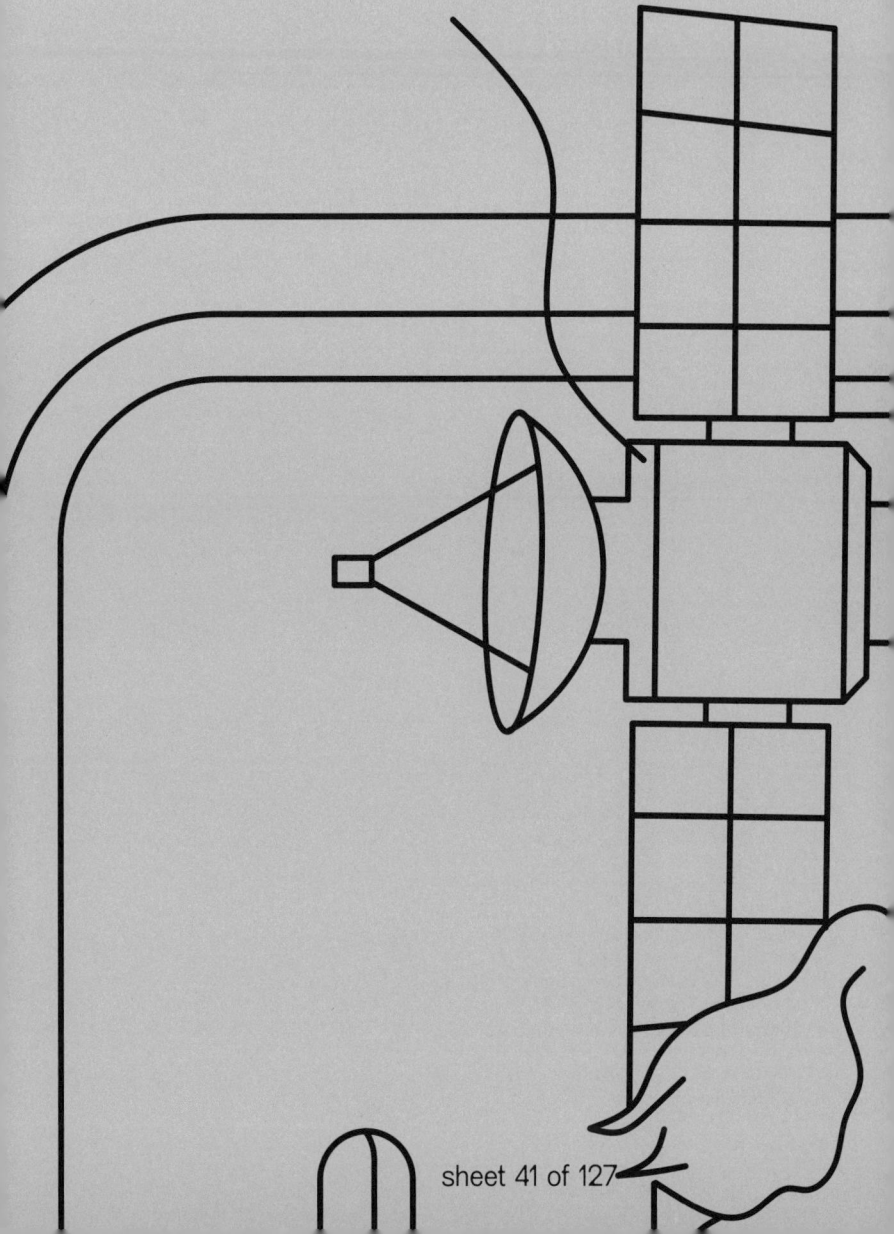

We *need* privacy-preserving protocols that can serve the interests of civil society and small organisations that cannot afford the massive infrastructure bills associated with CDN corporations. We *need* privacy on the internet to be accessible to everyone. Privacy is a political right. It cannot and should not be a premium service.

58 Notes:

52. Cloudflare's ODoH announcement: https://blog.cloudflare.com/oblivious-dns/
53. Chrome partnered with Fastly for OHTTP: https://developer.chrome.com/blog/oblivious-http-for-k-anon-server-with-fastly/
54. Private Relay with Fastly, Cloudflare and Akamai partners: https://www.streamingmediablog.com/2021/06/apple-private-relay.html
55. https://letsencrypt.org/2016/09/20/what-it-costs-to-run-lets-encrypt.html
56. ISRG's product around privacy-preserving measurement: https://divviup.org/
57. 'Designing cryptography for small organisations and projects' talk under Crypto for People session at https://rwc.iacr.org/2023/program.php

59 Confidentiality washing in online advertising

60 Michael Veale   Michael Veale is associate professor in the Faculty of Laws, UCL. He researches tensions between emerging technologies, their societal impacts, and their legal characteristics.

61 Keywords: big tech, confidentiality, encryption, online ad economy, privacy

62   Online advertising has long been a privacy shitshow. Whenever apps or websites try to show an ad to someone, they send data about that viewer to hundreds of would-be bidders, who act for big brands looking for the right audiences. These bidders in turn send the data to hundreds more data brokers, to learn as much as possible about the viewer—all just to work out a reasonable maximum bid for their attention. This arrangement has long been deeply illegal, but its complexity and opacity has left regulators paralysed 〰. Thanks to new(ish)

63 laws, regulators creaking into action, and tracker blockers in browsers, we may see the end of this invasive practice.

It would, however, be naïve to think this will be the end of online tracking. Advertising has left so many marks on internet economics and internet infrastructures that it can be a struggle to imagine alternatives. In this piece, we look at the power struggles emerging around these replacements, as well as examining what exactly these proposals improve upon, and which aspects might instead be spin, distraction, or mirage.

Proposed replacements centre on a bundle of intriguing tools called 'privacy-enhancing technologies', or PETs. To use PETs, first you think of a task you want to do that typically relies on a lot of people moving a lot of data— like browsing the Web, analysing medical records from many clinics, or communicating with a group of people—and then build a set of technologies that lets you do it while only collecting or disclosing the minimum of information to untrusted people 〰. The results can be surprising and counterintuitive. In the case of online advertising, imagine deeply targeted ads, but chosen and delivered without personal data ever leaving your device in a form others can pry upon. PETs excite people precisely because it appears that society can both have its cake and eat it: keeping the business model, while protecting privacy.

64

PRIVACY IS POWER—BUT NOT FOR YOU

PETs seem like a win-win situation, but not everyone wins when you build encrypted systems. PETs generally require readable data to stay on people's devices, rather than leave to servers. Anyone can run a server, but a small handful of firms have real strangleholds on the large-scale computational infrastructures needed for PETs, including for what devices like phones can do. These include operating system providers—like Apple and Google—browser providers—like Apple and Google—and app store providers like—yes!—Apple and Google. In contrast, smaller adtech firms are not well positioned to benefit from a change to PETs. Such adtech firms' infrastructure is only skin-deep; it cannot do complex computing, as is required for most PETs to function.

PETs even make a company like Meta feel vulnerable. It controls significant parts of the top of the technology stack, such as apps, and even some of the bottom, like undersea cables but lacks much of the middle. Meta has no major operating system, browser, or app store. This leaves the firm reliant on decisions made by Apple and Google. When Apple forced an opt-in function for a certain type of tracking across apps in mid-2021, Meta reported a significant hit to its revenue, as this limited the firm's ability to gather data on which apps people use and how. The firm is undoubtedly worried more moves are yet to come ⌁.

65 →

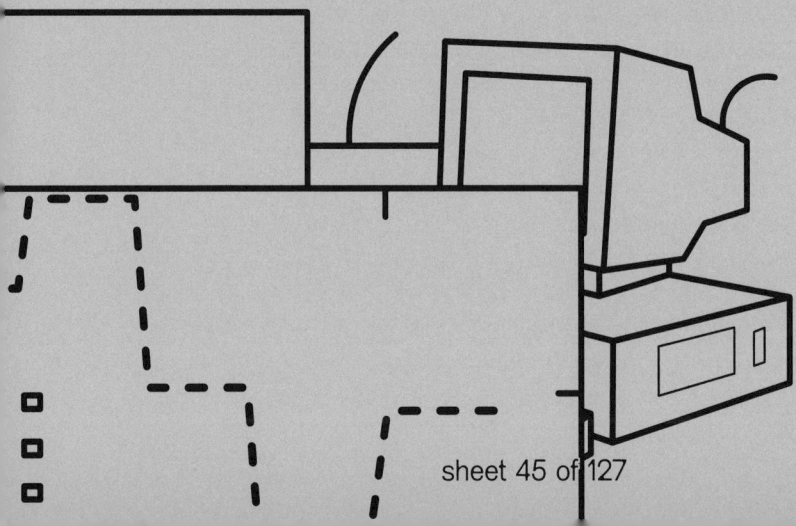

CONFIDENTIALITY WASHING

Data giant Meta's PET anxieties show the power
PETs can give infrastructure providers but will
probably not stir huge sympathies in the reader. Yet
there are other reasons to keep a critical eye over
the specific way PETs are developing. If PETs are
a solution to adtech's woes, what is the problem?
Proponents would have you believe the problem
is a lack of confidentiality—the ability to protect
data about you from being looked at by others.
This is certainly a problem. The leaky adtech stack
is regularly abused in diverse ways, from outing
individuals' sexual orientation to being piggybacked
upon by intelligence services and the military to
identify targets ⌇⌇⌇. But what problems remain   ⟵   66
even when confidentiality is secured?

We can imagine an on-device, targeting system
which transfers no data off-device, yet bases
ads on comprehensive browsing data or deeply
personal health data from wearables. This may
be confidential, but is it private? Even if a tech
company employee cannot read your Web history,
your device itself is trampling over personal
boundaries in ways you may not be able to stop
or control. Even if you could turn these settings off,
we can imagine a company making access to free
services, like cloud storage or online subscriptions,
conditional on confidential analysis and targeting.
Because this data is kept confidential, there is even
a potential perverse outcome where companies try
to use more sensitive data than before as part of
their business models, arguing that it's fair game
if not transmitted or centralised. This would be a
world of confidentiality washing, where devices
and their corporate manufacturers, instead of the

67 → servers of platform companies, are betraying, profiling, and manipulating users—under the guise of confidentiality 〰〰. Is that much of an improvement?

68 ⌐

ENCRYPT ALL THE THINGS?

This situation places civil society organisations in a bind. Since Snowden, they have successfully rallied companies and the public behind encrypting communications, limiting some forms of state surveillance. But encryption technologies have moved on, and they can now more effectively encrypt analysis and computation as well. This is a much more open design space, where businesses can design complex PETs to advantage them to the detriment of their competitors 〰〰.

The debates about the politics of encryption are changing. It is no longer about encrypting a chat or a call. In many ways, it is more like encrypting a clinical trial, a company's tax payments, or a government policy. Activists defend encryption by talking about freedom to decide how we communicate—the means—regardless of what messages contain. Critics respond by reframing it as about the ends of certain communication, typically highlighting child abuse and serious crime. But encryption is no longer just about communication or expression. The ends that can be encrypted now include large-scale data analysis or entire internet business models. This quickly, and confusingly, blurs personal privacy and corporate opacity.

In an already highly technical domain, the thought of further complicating messaging will make civil society queasy. But unless they start to navigate these turbulent waters, they risk tying the legitimate protection of encrypted expression to the questionable legitimacy of any business or surveillance practice that can be confidentiality-washed using PETs. As encryption goes from a narrow-purposed set of tools to an infrastructure supporting broad, open-ended computing systems, we cannot afford to see it as simply 'good' or 'bad'. Instead, we must ask the questions we should always be asking of all powerful systems: Who put them there? Who do they functionally benefit? And most importantly—how can we negotiate, or if necessary, refuse and reject them?

---

69 Notes:

63. Veale, M. and Borgesius, F.Z. 2022. Adtech and Real-Time Bidding under European Data Protection Law. *German Law Journal* 23: 226–256 doi.org/jphn; Veale, M., Nouwens, M. and Santos, C. 2022. Impossible Asks: Can the Transparency and Consent Framework Ever Authorise Real-Time Bidding After the Belgian DPA Decision? *Technology and Regulation* 2022, 12–22. doi.org/gpgcfb

64. Gürses, S., Troncoso, C. and Diaz, C. 2015. Engineering Privacy by Design Reloaded. Amsterdam Privacy Conference 2015. https://perma.cc/C77G-5EP9

65. Meta has thus been active within internet standardisation bodies, trying to shift the narrative towards PETs which rely more on servers—so far with limited success. See archived Google Slides at https://perma.cc/LBW3-TB64; archived Google Doc at https://perma.cc/28UN-F8VB; Veale, M. 2022. Future of online advertising: Adtech's new clothes might redefine privacy more than they reform profiling. netzpolitik.org. https://perma.cc/EYC4-66H6

66. Modderkolk, H. 2014. Lees hier hoe de Britse geheime dienst GCHQ Belgacom aanviel. NRC. https://www.nrc.nl/nieuws/2014/12/13/verantwoording-en-documenten-a1420301; Soltani, A., Peterson, A. and Gellman, B. 2013. NSA uses Google cookies to pinpoint targets for hacking. Washington Post. https://www.washingtonpost.com/news/the-switch/wp/2013/12/10/nsa-uses-google-cookies-to-pinpoint-targets-for-hacking/; O'Brien, M., and Bajak, F. 2021. Priest outed via Grindr app highlights rampant data tracking. AP. https://apnews.com/article/technology-europe-business-religion-data-privacy-97334ed1aca5bd363263c92f6de2caa2

67. See relatedly Berjon, R. 2021. The Fiduciary Duties of User Agents. Pre-print https://doi.org/10/gjw466; Renieris, E. 2021. Why PETs (privacy-enhancing technologies) may not always be our friends. Ada Lovelace Institute. https://perma.cc/E84R-V93V

68. Rogaway, P. 2015. The Moral Character of Cryptographic Work. Essay accompanying the IACR Distinguished Lecture, AsiaCrypt 2015. https://perma.cc/KV9S-B7LJ

70 Mining 'Silicon Holler' for alternatives

---

71 Britt Paris

Britt S. Paris, Assistant Professor at Rutgers University, hails from Missouri coal country.

72 Keywords: blockchain, data centres, extraction, public utility internet, resistance

---

73 Eastern Kentucky is a site of ongoing overlapping political economic and environmental crises. These crises compel interested parties to continually reimagine infrastructure. This chapter traces the throughlines between the potential of the (Silicon) Valleys and the 'Hollers'—low places between hills or mountains—of rural Appalachian America. Two visions for the future of internet infrastructure in Eastern Kentucky—one rooted in blockchain and one in cooperative networks—implicate different aspects of the region's past, and signal different possibilities for the future.

THE PROMISE AND PITFALLS OF BLOCKCHAIN

One infrastructural vision for Kentucky hinges on 'Blockware Solutions'. This cryptocurrency mining and infrastructure company is privately owned and operated by individuals far outside the state. It announced in 2022 that it built a 20-megawatt crypto mining centre in an old mining office. As such, it is a site that directly recalls Kentucky's coal industry, in several ways. These mining facilities run blockchain's digital ledger checksums to generate cryptocurrency. These blockchain tasks require massive amounts of electricity to support computing power and data storage. Blockware Solutions rents absentee customers, 'co-located' space in the centre instead of owning and running the physical servers, reducing capital costs to Blockware's customers. Extractive development is a familiar promise in Kentucky, where miners have long hoped to strike it rich and privatised companies benefit from booms.

These private blockchain solutions demand public resources. Co-located crypto mining facilities in Kentucky reap the benefits of operating in an area that has low-cost electric power 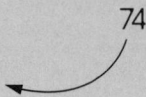. The lower electricity costs exist primarily because of the predominantly smaller public electricity utilities—often member-run cooperative electric utilities—offering low prices to a poorer population. Blockware Solutions runs on these same electric utilities and on the public utility internet backbone, the Eastern Kentucky Network . For a lesser part, the cheaper electricity is due to utilities' proximity and use of coal, which powers much of the electricity provision. Despite Blockware's claims, the energy required for the facility is not 'clean'; Kentucky is the highest carbon-emitting state in the Bitcoin network. Blockware Solutions thus profits from mining public resources in the 'Hollers', extending environmental harms.

74

75

76

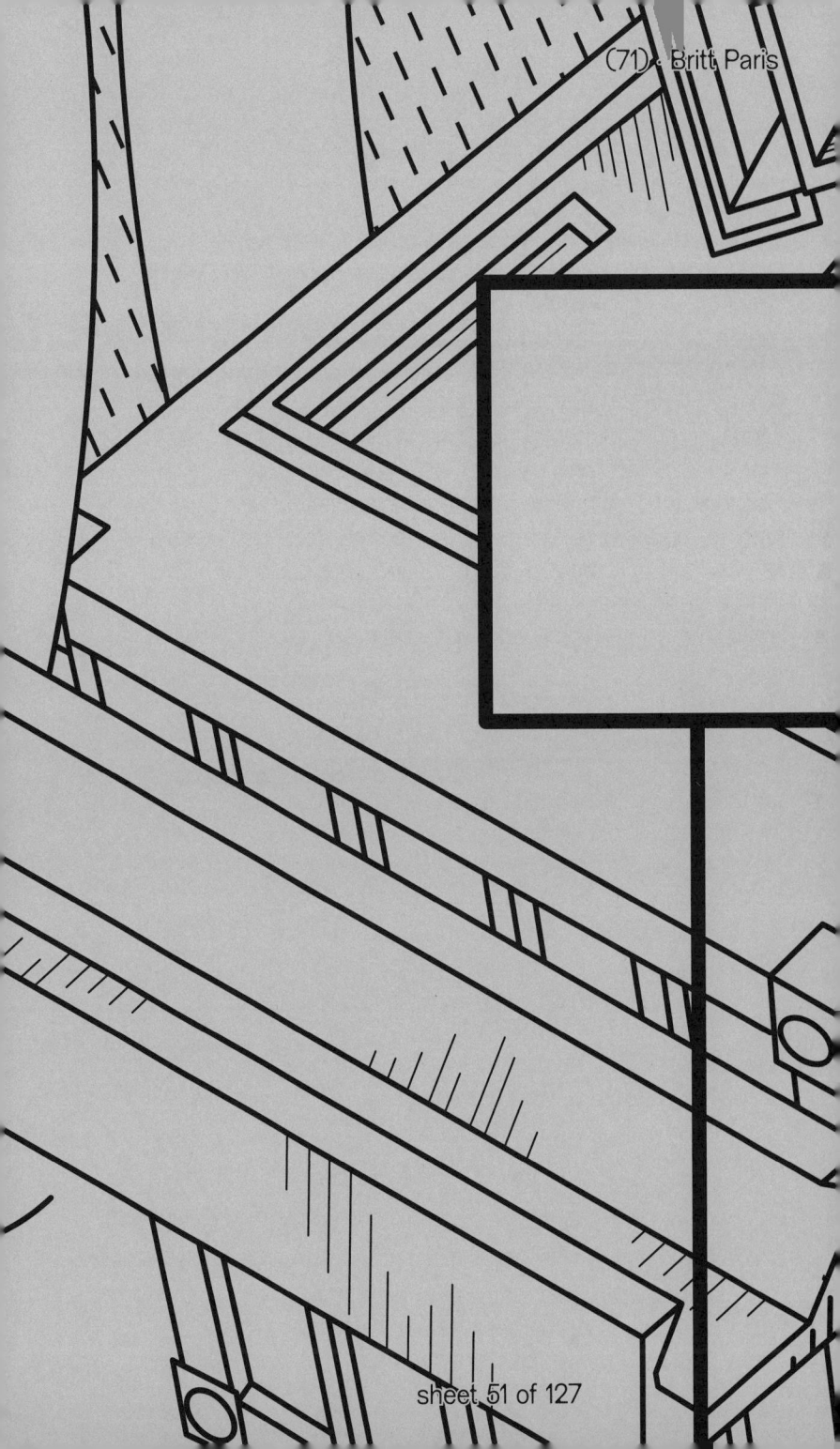

The extraction of public goods for private profits is long familiar to the region, and crucially, supported by its politicians. The collocated, data-centric vision of internet infrastructure in Kentucky is especially popular among the state's public officials. Lawmakers from both parties have welcomed cryptocurrency mining operations. In 2021, Democratic Governor Andy Beshear signed Kentucky house and senate bills that respectively offer crypto miners a series of sales tax breaks. Republican Senators Wil Schroder and Brandon Smith had investments in cryptocurrencies until they entered bill sponsorship that caused conflicts of interest. They remain publicly enthusiastic about the job prospects that come with cryptocurrency mining.

77

78, 79

However, the promises of economic development and new jobs remain unmet. The Blockware Solutions plant will only employ seven full-time workers. Some Kentuckians would rather tax out-of-state corporations to fund some of Kentucky's many needs. The collocated data centre, the blockchain solution, opens onto an exploitable future that echoes the past.

80

COOPERATIVE, PUBLIC UTILITY INTERNET

Eastern Kentucky has a widely-unknown history of radical politics, and enjoys more radical visions for infrastructural development. McKee, Kentucky is the county seat of one of the most economically depressed counties in the state (US Census Bureau 2021). However, its co-operative telecommunications company, the People's Rural Telecommunications Cooperative (PRTC), offers free high-speed fibre-optic internet installation and

81

82, 83 → service to local schools, hospitals, libraries and even some households ⬡.

The PRTC's past stretches into the future. PRTC's history dates to the New Deal's Tennessee Valley Authority, which brought electricity to rural places across the US, later followed by New Deal public works programmes for rural telephony. The PRTC runs the same as it has for over 70 years. The member-elected board meets monthly with members to discuss needs and vote on new projects, pricing, and a host of other activities. Whatever is not used to pay off government loans or for service provision, is returned to the members in the form of yearly checks issued by PRTC.

The PRTC provides an alternative model for the internet's future where people, not corporate profits, are centred. Because of the enhanced internet service offered by the PRTC and

84, 85 → other cooperatives, some have started to call the area 'Silicon Holler'. The term was coined by Republican congressman Hal Rogers ⬡ , to market rural Appalachia and its 'Hollers' as tech hubs full of workers desiring jobs. Democrat congressman Ro Khanna from Silicon Valley

86 → agrees with such posturing ⬡. But, much of the activity in Silicon Holler attracts outside venture capitalists, and companies like Blockware Solutions looking for cheap public infrastructure to exploit, entrenching an extractive politics through internet infrastructure development.

CENTRING PEOPLE

These alternative internet solutions intertwined in the Hollers of Eastern Kentucky are both

examples of what happens when residents of
an area experience years of overlapping crises
and struggle towards stability. Lawmakers
intervene by allowing the privatisation of
public infrastructure predicated on the
empty promise of much-needed
economic resources, like

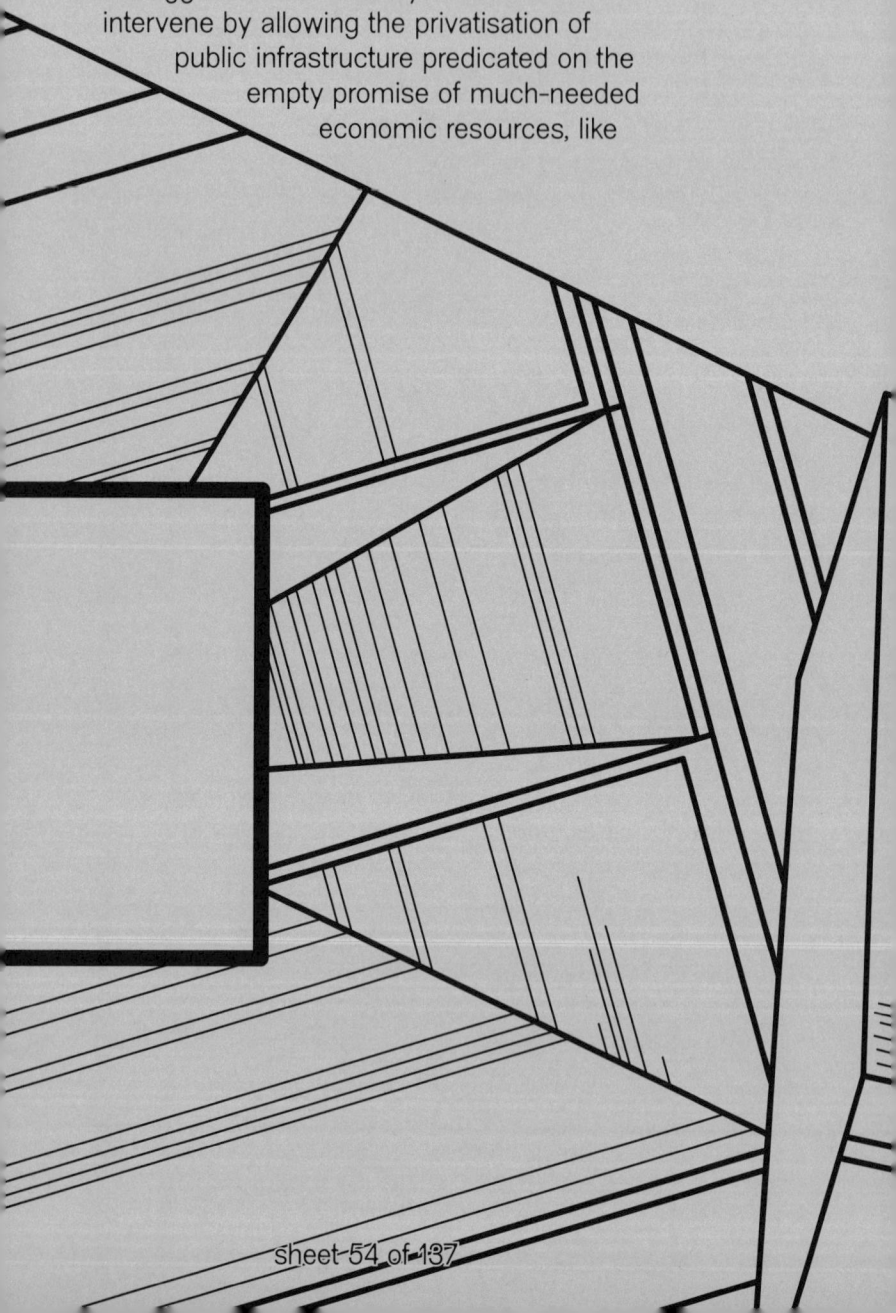

jobs, as in the case of Blockware Solutions. On the other hand, the PRTC shows how inhabitants in the area organise and maintain public cooperative infrastructure that benefits them. The PRTC reminds us that internet service provision is deployed and governed locally, then connects to the rest of the world. As such, the PRTC's cooperative model presents an example to draw that can guide organising broad coalitions towards imagining and building a people-centred internet.

---

87 Notes:

74. US Chamber of Commerce, Average Electricity Retail Prices. Global Energy Institute, 2020. https://www. globalenergyinstitute.org/average-electricity-retail-prices-map

75. Eastern Kentucky Network History, Eastern Kentucky Network, 2022. https://www.ekn.com/#history

76. de Vries, A., Gallersdörfer, U., Klaaßen, L. and Stoll, C. 2022. Revisiting Bitcoin's carbon footprint. *Joule* 6 (3) 498–502. doi: https://doi.org/10.1016/j.joule.2022.02.005

77. Nelson, D. 2021. Kentucky Governor Signs Tax Breaks for Crypto Miners Into Law. *Coindesk.* https://www.coindesk.com/markets/2021/03/26/kentucky-governor-signs-tax-breaks-for-crypto-miners-into-law

78. Kentucky Legislative Ethics Commission, Statement of Financial Disclosure. State of Kentucky, Feb. 15, 2022.

79. Estep, B. 2022. Kentucky's digital gold rush. What's behind the crypto mining boom in coal country? *Lexington Herald Leader,* Apr. 21, 2022.

80. Bailey, J. 2021. Governor Should Veto Tax Bills That Reduce Needed Revenues, Put Kentucky at Risk of Owing Back Aid. Kentucky Center for Economic Policy, Mar. 23, 2021. https://kypolicy.org/governor-should-veto-tax-bills-that-reduce-needed-revenues

81. Hennen, J.C. 2008. *Harlan Miners Speak: Report on Terrorism in the Kentucky Coal Fields.* Lexington: University Press of Kentucky.

82. Halpern, S. 2019. The One-Traffic-Light Town with Some of the Fastest Internet in the U.S. *The New Yorker.* https://www.newyorker.com/tech/annals-of-technology/the-one-traffic-light-town-with-some-of-the-fastest-internet-in-the-us

83. Ali, C. 2021. *Farm Fresh Broadband: The Politics of Rural Connectivity.* Cambridge, MA: The MIT Press.

84. Rogers, H. 2017. Connecting Silicon Valley to "Silicon Holler". https://halrogers.house.gov/2017/3/connecting-silicon-valley-to-silicon-holler

85. Rosenblum, C. 2017. Hillbillies who code: the former miners out to put Kentucky on the tech map. *The Guardian.* https://www.theguardian.com/us-news/2017/apr/21/tech-industry-coding-kentucky-hillbillies

86. Khanna, R. 2022. What Silicon Valley Can Learn From 'Silicon Holler'. *The Atlantic.* https://www.theatlantic.com/ideas/archive/2022/02/ro-khanna-digital-revolution-silicon-valley-jobs/621421

88 The problem is growth

89 environmental harms of tech revisited

90 Fieke Jansen

Fieke Jansen is the co-founder of the critical infrastructure lab and a Post-doctoral Researcher at the University of Amsterdam. She is also the coordinator of the Green Screen climate justice and digital rights coalition.

91 Keywords: data centres, economics, environmentalism, grassroots organising

92 Pollution in lithium mines in the Andes region and e-waste dumps in Ghana feel like worlds apart from the sterile data centres that store our data. Yet, the latter can't operate without the former. The internet's policy, engineering and governance communities are slowly coming to grips with the environmental harms of technology, but the sustainability solutions that circulate in response are often narrow in scope. In the fight against climate change and environmental degradation, we need to deepen, rather than flatten, the imaginary of what the problems and solutions are.

Indeed, the current notion of developing 'sustainable infrastructures' neglects the uncomfortable reality that economic growth, rather than inefficiency, might be the problem. As such, narratives and actions that seek to optimise single issues, like energy consumption in data centres, offer misleading solutions as these

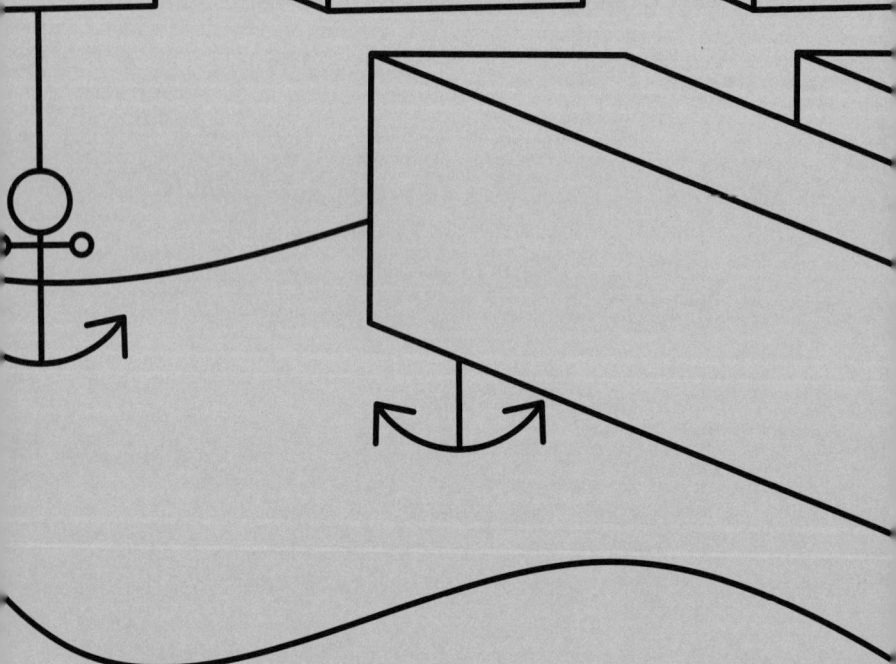

perpetuate extractive growth models on which our data economy is based. Critical approaches to sustainable internet infrastructures, which engage with solidarity, reduction, limits, and redistribution, are needed.

THE ENVIRONMENTAL IMPACT
OF THE INTERNET

93

Internet infrastructures contribute to carbon emissions and the extraction of natural resources. Despite the 'cyberspace' framing, these technologies are deeply implicated in conflicts over land, water, and energy consumption △. Critically

attending to infrastructure allows us to see
which politics are embedded, or excluded, from
responses to the internet's environmental harms.
Some scholars have quantified the environmental
impact of the extraction and production of raw
materials, such as silicon and lithium, that are
94   needed to manufacture end-user devices or
expand internet infrastructure △. Others have
calculated the energy consumption, carbon

emissions, and water usage of digital tasks such as internet searches, training Artificial Intelligence (AI) models, and the operation of data centres ◁. This research points to the many ways in which internet infrastructures are entangled with questions around extraction and pollution.

95

These developments need to be placed against a backdrop of infinite growth, scalability, resource, and data extraction of internet infrastructures. A central ideology of the internet industry is that when networks grow, the cost for each new node increases linearly but the value of the network increases exponentially ◁. Thus, the value of infrastructures, and the services that run on top of it, are intrinsically connected to the market shares of Big Tech companies and the volume of users, nodes, and data they contribute. The operation, maintenance, and expansion of the internet infrastructures needed to accommodate this growth will involve further environmental harm.

96

POLICY AND INDUSTRY RESPONSES

Responses to the undeniable relationship between internet infrastructures and the climate crisis fall into two categories. In the first response, the tech industry tries to minimise the environmental impact of internet infrastructures by increasing the efficiency of data centres, transitioning to renewable energy sources to reduce dependency on the fossil fuel industry, and offsetting carbon emissions. The second response presents computation as the solution to pressing social problems, including the climate crisis and environmental degradation. Internet infrastructures are presented as a critical component in planning for and responding to

97 climate change by making other industries, such
as transportation and agriculture, more efficient △.
These twin responses are contentious for a
number of reasons that have been explained by
98 others △, but primarily, there is little evidence
that one can decouple business growth from its
inevitable environmental and climate impact.

More importantly, however, these responses
offer insights into the political and economic
interests that are shaping the future of our
societies and planet. Climate change and
environmental degradation are complex problems
that are intrinsically tied to capitalism and the
drive for perpetual economic growth. Narrow
sustainability solutions, however, each with well-
documented limitations, always allow individual
99 companies to ensure maximum value extraction
while externalising harms. This is the myth
of green capitalism △: the widely held belief
that the environmental problems caused by
economic growth can be solved with more but
different economic growth. To ensure that the
world remains habitable for current and future
generations, we need to be able to question the
presumed compatibility between economic growth
and the environment.

TOWARDS A MORE HOLISTIC APPROACH

Internet infrastructures have a demonstrated
impact on the environment, as a growing body of
research shows △. Policy and industry responses
100 to these harms are primarily aimed at mitigation,
however, and neglect to address the uncomfortable
reality that economic growth itself might be the
problem. Any critique raises the question of what

climate-supportive internet infrastructures could actually look like. There are no easy answers to this question. As a community of internet and sustainability practitioners, we are beginning to wrap our heads around the complexity of the problem and starting to articulate what sustainable and equitable infrastructures could look like. It is early days, but deeper approaches start with building alternative models, acting in solidarity with people and the planet, and exploring the values of reduction, limits, and redistribution.

There is an emerging community of grassroots organisers, environment and digital rights groups, academics, public interest technologists and others, who organise around the environmental implication of technology. This community challenges the dominant framing of 'green tech' and is discussing what internet infrastructures that centre sustainability over perpetual economic growth might look like. Initiatives range from Branch magazine △, an online magazine written by and for people who dream of a sustainable and just internet for all. Low-tech magazine △, which refuses to assume that every problem has a high-tech solution. And the solar protocol △, a hosting proposition that routes internet traffic across a network of solar-powered servers. Websites are loaded from where the sun shines. The value of these initiatives lies in their ability to challenge the norm of green capitalism and create an alternative infrastructural imaginary.

101

102

103

104 Notes:

93. For example, in 2021 Google emitted over 47 million tonnes of CO2 and used over 23,836 million litres of water. This means that the company exceeded the environmental impact of the entire country of Laos. See: Robinson, S., Hellstern, R. and Diaz, M. 2022. Sea Change: Prioritizing the Environment in Internet Architecture. IAB workshop on Environmental Impact of Internet Applications and Systems, 2022.

94. Sutherland, B. 2022. Strategies for Degrowth Computing. Eighth Workshop on Computing within Limits 2022. https://doi.org/10.21428/bf6fb269.04676652. Williams, E.D. 2004. Environmental impacts of microchip manufacture. *Thin Solid Films* 461 (1) 2–6. https://doi.org/10.1016/j.tsf.2004.02.049

95. Hao, K. 2019. Training a single AI model can emit as much carbon as five cars in their lifetimes. MIT Technology Review. https://www.technologyreview.com/2019/06/06/239031/training-a-single-ai-model-can-emit-as-much-carbon-as-five-cars-in-their-lifetimes/; Bender, E.M. et al. 2021. On the Dangers of Stochastic Parrots: Can Language Models Be Too Big? Proceedings of the 2021 ACM Conference on Fairness, Accountability, and Transparency, pp. 610–623. https://doi.org/10.1145/3442188.3445922; Siddik, M.A.B., Shehabi, A. and Marston, L. 2021. The environmental footprint of data centers in the United States. *Environmental Research Letters* 16 (6) 064017. https://doi.org/10.1088/1748-9326/abfba1

96. Metcalfe, B. 2013. Metcalfe's Law after 40 Years of Ethernet. *Computer* 46 (12) 26-31. doi: https://doi.org/10.1109/MC.2013.374

97. European Commission. 2020. White Paper on Artificial Intelligence - A European Approach to Excellence and Trust. Brussels: European Commission. https://ec.europa.eu/info/sites/default/files/commission-white-paper-artificial-intelligence-feb2020_en.pdf; Pargman, D. et al. 2020. From Moore's Law to the Carbon Law. Proceedings of the 7th International Conference on ICT for Sustainability. New York, NY: Association for Computing Machinery (ICT4S2020), pp. 285–293. https://doi.org/10.1145/3401335.3401825; Panel at European Parliament on A Digital and Green Transition: Will Artificial Intelligence foster or hamper the Green New Deal? 3 February 2021.

98. Day, T., Mooldijk, S., Smit, S., Posada, E., Hans, F., Fearnehough, H., Kachi, A., Warnecke, C., Kuramochi, T. and Höhne, N. 2022. Corporate Climate Responsibility Monitor 2022: Assessing the transparency and integrity of companies' emission reduction and net-zero targets. NewClimate Institute. https://newclimate.org/sites/default/files/2022/02/CorporateClimateResponsibilityMonitor2022.pdf; Greenfield, P. 2023. Revealed: more than 90% of rainforest carbon offsets by biggest certifier are worthless, analysis shows. https://www.theguardian.com/environment/2023/jan/18/revealed-forest-carbon-offsets-biggest-provider-worthless-verra-aoe

99. Buller, A. 2022. The value of a whale. Manchester University Press; Pistor, K. 2021. The Myth of Green Capitalism. Project Syndicate. https://www.project-syndicate.org/commentary/green-capitalism-myth-no-market-solution-to-climate-change-by-katharina-pistor-2021-09

100. Ariadne. 2022. Funding at the intersection of Climate and Tech. https://www.ariadne-network.eu/funding-at-the-intersection-of-climate-and-tech

101. https://branch.climateaction.tech

102. https://www.lowtechmagazine.com

103. http://solarprotocol.net

PART 2

GOVERNMENT CAPTURE OF

INTERNET
INFRA-
STRUCTURE

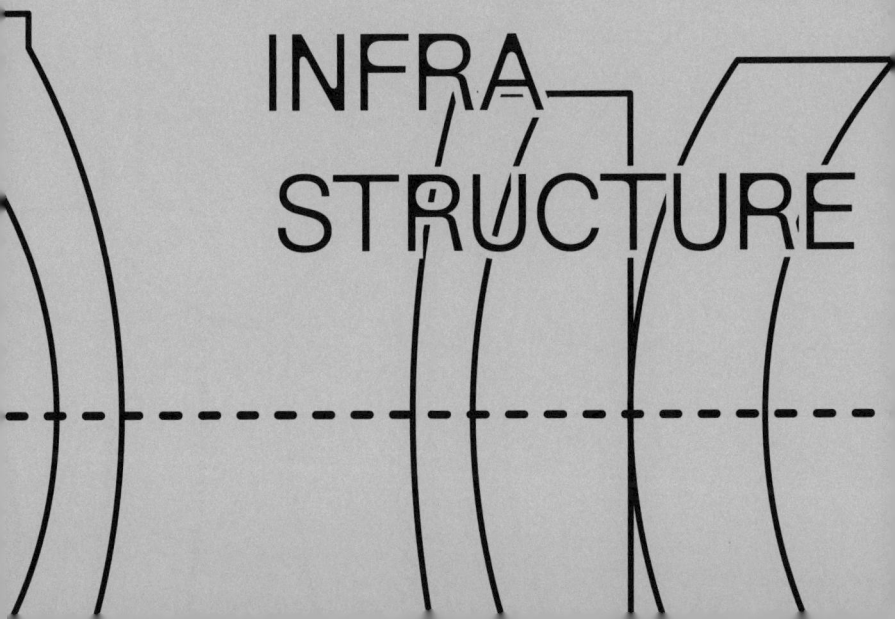

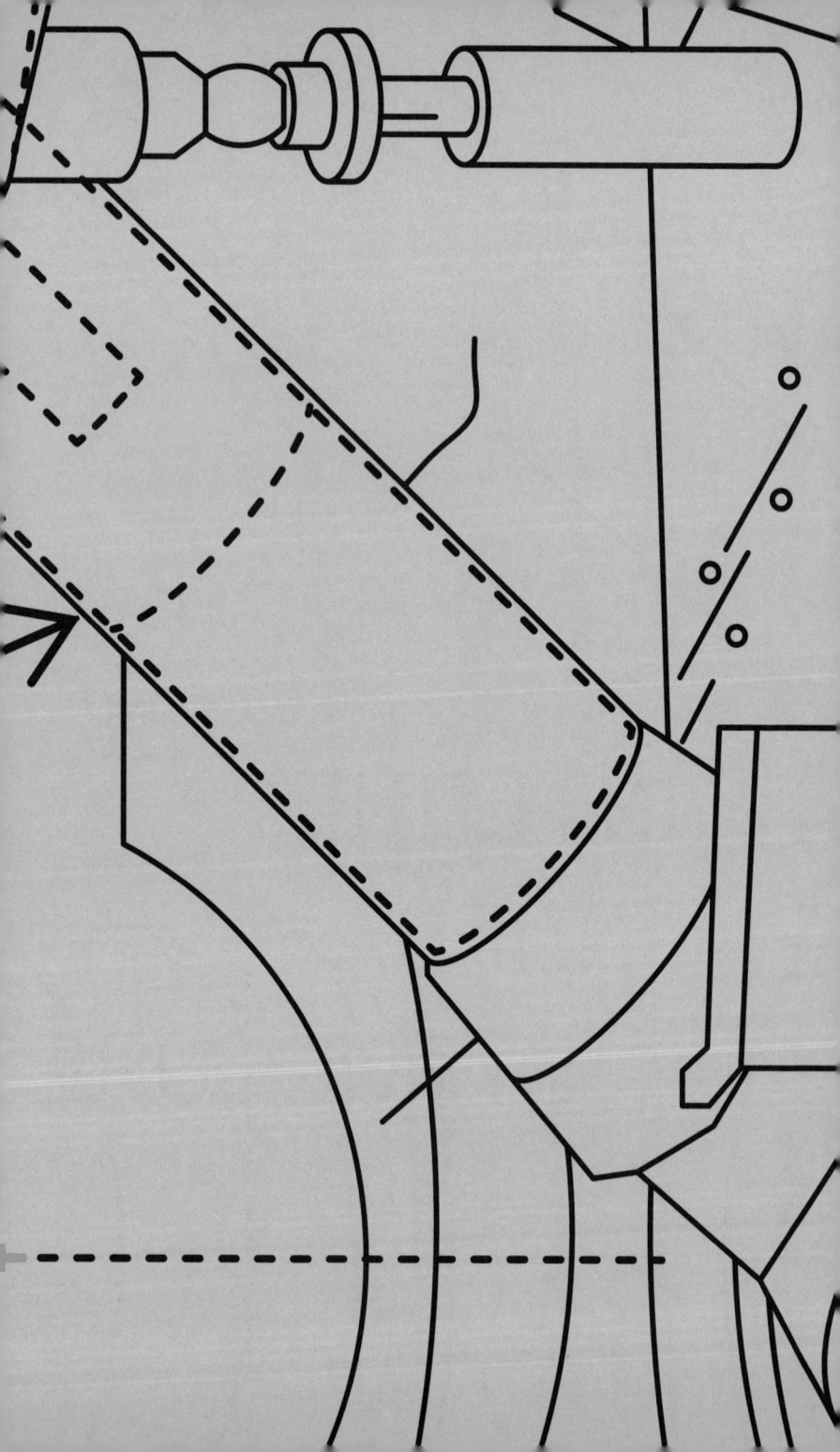

107 the colonial religious history behind online hate

108 Suzanne van Geuns   Suzanne van Geuns is writing a genealogy of sexual frustration on the internet. She works at Princeton's Centre for Culture, Society, and Religion.

109 Keywords: 8chan, Asia, Cloudflare, colonialism, hate speech

110 'There's really no other way to put it,' Frederick Brennan sighed when his interviewer pushed him to consider the hate spread through his website,

111 'than that we're a common carrier' 📓 . Brennan established 8chan in 2013, and the message board quickly became notorious for its prominent

112 role in harassment campaigns, white nationalist mass murders, and anti-democratic conspiracy movements. Speaking with *Ars Technica*, he explained that he saw 8chan as a 'pipeline' for content and rushed to underline that only 'some bad people' among its many users used this pipeline for hate. When Brennan refused to plug it, however, even after multiple violent events directly associated with 8chan, infrastructure corporations intervened. They refused the website hosting services or website protection, cutting access to their infrastructural pipelines to stop the flow of hate through 8chan 📓 .

This concerted effort to 'clog' should not trick us into thinking that justice is a pipeline matter, however. Cloudflare, a company that removed 8chan from its client base, explained its decision in language that is ironically similar to Brennan's. Cloudflare provides 'mere conduits' for content, and hastened to underline that its specialised Content Delivery Networks principally service the 'huge portion of the Internet' that is not dedicated to hate 📖. It is as though to invoke pipelines is to reduce the politics of hate to a yes/no question about flow and blockage.

113

The strange case of 8chan is also a chance to probe deeper, however. An American political menace, 8chan was powered from the Philippines and paid for with Japanese money. How did this come to be?

THE PHILIPPINES

'Hold on a sec,' Brennan tells his interviewer, 'I think we missed a few calls from PLBC.' He was discussing his average workday ('coding, trying to stop spam') running 8chan 📖. PLBC is a mishearing. Brennan almost certainly said 'PLDT, short for Philippine Long Distance Telephone company. He was calling from the Philippines, where he moved to join his then-business partner Jim Watkins. Manila was affordable and pleasant, Watkins had promised, and the work of running a message board could be done from anywhere. The pipelines that connect computers to servers span the globe. Still, it is no coincidence that the duo ended up in the Philippines.

114

PLDT's physical infrastructure was born of imperialist ambition. Its establishment was shepherded by American statesman Henry Stimson, who was then working as Governor-General of the Philippines 📖. He was committed

115

116 to a version of colonialism that presented itself not as extraction but as benevolent 'tutelage' ▩. Drawing on a ubiquitous form of Christian racism, in which Filipino people were understood to be in an earlier stage of evolutionary and religious development than white American Christians, Stimson and his peers set out to 'uplift' the Philippines and 'prepare' its inhabitants for self-

117 governance ▩. They expected to facilitate this uplift in part through the 'material development' of

118 Philippine infrastructure ▩.

Colonial infrastructure was a capitalist enterprise. US colonialists believed trade would mould the Filipino into the 'right kind of self-possessed

119 individual,' a subject whose fitness for self-governance is indexed by her ability and desire to consume American products ▩. Communications infrastructure, which presented opportunities for advertising *and* for Christian missions, seemed

120 especially essential for setting people in the Philippines on the American path ▩. Once the right pipelines were in place, men like Stimson reasoned, the right kind of civilisation would soon develop at the other end. Their ideological commitment to infrastructure, as part of the slow realisation of imperial and capitalist politics, still undergirds physical internet infrastructures today.

PLDT is a case in point. In the early twentieth century, when the United States first gained control over the Philippines, there was a rush to ensure that the ocean cables required for telephone conversations would be laid by *American*

121 businesses ▩. The undersea cables supporting the Web still follow these routes today, and as a result, the Philippines is particularly well-positioned

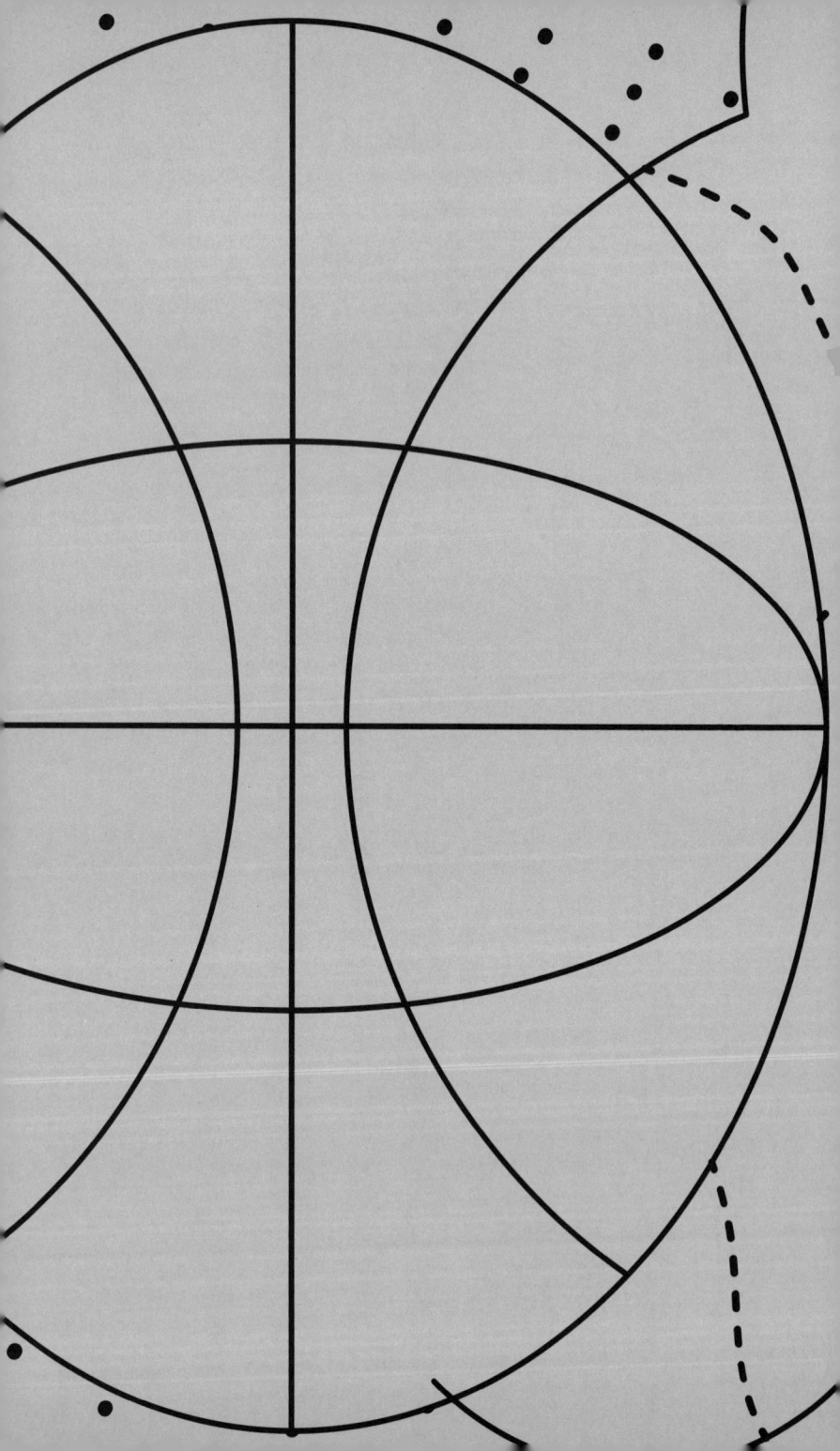

122 → in this submarine network 🖥️. PLDT, the company in which these initial American infrastructural investments would culminate, enjoyed a long monopoly as an internet service provider 🖥️. It still

123 → offers the lowest latency in the region 🖥️ PLDT

124 → would have been the obvious choice for high-volume users such as Brennan, the practical and seemingly neutral usefulness of its 'mere conduits' inseparable from colonial ideologies.

THE FLOW OF PROFIT

While the pipelines between Asia and the United States may have been built to facilitate 'progress' toward white Christian superiority, their immediate function was always to facilitate the flow of capital. When Brennan flew to Manila, gladly accepting his offer of an apartment and infrastructural services, he joined an experienced exploiter of such flows. Watkins is a US Army veteran turned internet entrepreneur. His profit model was to host content

125 → in the United States—through a server farm in San Francisco, to be precise—that would be censored in the regions where it is most popular 🖥️. He hosted porn. N.T. Technology, Watkins's main business venture, used US-based servers to turn a lack of constraints around speech, an archetypal feature of the American legal landscape, into an opportunity for financial gain.

Keeping 8chan running, it turned out, required exactly this form of global exchange. In the late 1990s, N.T. Technology began hosting Japanese porn with the genitals unblurred. In the early

126 → 2000s, the company expanded to host the popular and controversial Japanese discussion board 2chan 🖥️. The board was profitable. In

2008, its owner boasted a million dollars a year in advertising revenue . When asked whether he, as owner, should be held responsible for the harassment campaigns spawned on 2chan, he spoke of pipelines—would a cell phone carrier be responsible for every threatening phone call?—and confidently asserted that he was not afraid of crackdowns from the Japanese government, because 2chan was hosted in the United States. This confidence came at a steep price: Watkins claimed 60% of all 2chan's advertising proceeds .

127

128

Without this money, extracted from the intersection between Asian demand and American legal and material infrastructures, 8chan could not have acquired its cultural prominence. Almost from the start, 8chan— anarchic, unmoderated— was popular enough to need serious infrastructural support, but too offensive to monetise . Watkins's solution was simple: 8chan may have repelled advertisers, but 2chan did not. He used the latter to fund the former, perhaps in anticipation of 8chan eventually becoming as profitable as 2chan. Brennan's move to Manila interlaced the colonial pipelines that connect the

129

Philippines to the United States with the financial pipelines through which Watkins funnelled money from a profitable Asian messaging board to its not-yet-profitable American counterpart.

MAKING THE MOST OF IRONY

The story of 8chan is full of ironies, moments of ideological inconsistency and narrative instability. Consider the profitable export of American versions of freedom—whether the pleasures of consumption or those of porn—to racialised others in Asia, while American versions of hate are funded by the advertising proceeds associated with Asian media consumption. Or take infrastructures that were built from claims to moral and religious superiority, only to be re-used by men who actively dismiss moral questions by calling their work infrastructural and therefore beyond scrutiny.

Too often, debates about online hate smooth over such ironies as distractions from the all-important decision to plug or reroute a particular pipeline. But this is a missed opportunity for resistance; irony provides a foothold for a critique that is deeper than deplatforming. If we want different futures, we must uncover the submerged pipelines that tether today's cruelty to the colonial past—so that we can blow them up.

130 Notes:

111. https://arstechnica.com/information-technology/2015/03/full-transcript-ars-interviews-8chan-founder-fredrick-brennan

112. https://www.bellingcat.com/news/2019/11/04/the-state-of-california-could-have-stopped-8chan-it-didnt and https://arstechnica.com/information-technology/2019/11/breaking-the-law-how-8chan-or-8kun-got-briefly-back-online

113. https://blog.cloudflare.com/terminating-service-for-8chan

114. https://arstechnica.com/information-technology/2015/03/full-transcript-ars-interviews-8chan-founder-fredrick-brennan

115. Kalaw, M. 1928. Governor Stimson in the Philippines. *Foreign Affairs* 7: 372–83.

116. McKenna. 2017. American Imperial Pastoral: The Architecture of US Colonialism in the Philippines. Chicago: University of Chicago Press, 112; Wesling, M. 2011. Empire's Proxy: American Literature and U.S. Imperialism in the Philippines. New York: New York University Press, 3–6, 165.

117. Kramer, P. 2006. The Blood of Government: Race, Empire, the United States, and the Philippines. Durham, NC: University of North Carolina Press; Clymer, K. 1980. Religion and American Imperialism: Methodist Missionaries in the Philippine Islands, 1899-1913. *Pacific Historical Review* 49 (1) 29–50; Lum, K.G. *Heathen: Religion and Race in American History.* Cambridge, MA: Harvard University Press.

118. Wesling, note 6, at 10; McKenna, note 6, at 122; Kramer, note 7, at 201, 319.

119. McKenna, note 6, at 113.

120. Skelchy, R. 2020. The Afterlife of Colonial Radio in Christian Missionary Broadcasting of the Philippines. *South East Asia Research* 28 (3) 346–51.

121. Müller-Pohl, S. 2013. Working the Nation State: Submarine Cable Actors, Cable Transnationalism and the Governance of the Global Media System, 1858–1914. In Löhr, I. and Wenzlhuemer , R. (eds) The Nation State and Beyond: Governing Globalization Processes in the Nineteenth and Early Twentieth Centuries. Heidelberg: Springer Berlin, 116–17.

122. Starosielski, N. 2015. The Undersea Network. Durham, NC: Duke University Press, 40, 46, 177–78; Salac, R.A. and Kim, Y.S. 2016. A Study on the Internet Connectivity in The Philippines. *Asia-Pacific Journal of Business Review* 1 (1) 78–79.

123. Magtanong, O.O. 1997. Netting the Net: The Quest for Regulatory Mechanics for the Internet Service Sector in the Philippine Jurisdiction. *World Bulletin: Bulletin of the International Studies of the Philippines* 13 (5–6) 24–43; Serafica, R. 1998. Was PLDT a Natural Monopoly?: An Economic Analysis of Pre-Reform Philippine Telecoms. *Telecommunications Policy* 22 (4–5) 359–70.

124. Valeriano, C.M. et al. 2021. Application of Analytical Hierarchy Process in the Comparison of Internet Service Providers (PLDT, Globe, and Converge) in the Philippines. In *Proceedings of the International Conference on Industrial Engineering and Operations Management*, IEOM Society International, Rome, Italy, p. 1921.

125. https://www.washingtonpost.com/technology/2019/09/12/helicopter-repairman-leader-internets-darkest-reaches-life-times-chan-owner-jim-watkins

126. https://splinternews.com/meet-the-man-keeping-8chan-the-worlds-most-vile-websit-1793856249

127. https://www.wired.com/2008/05/mf-hiroyuki

128. https://www.wired.com/story/the-weird-dark-history-8chan Watkins later seized control of the 2chan domain, leading to an acrimonious business dispute—a manoeuvre he would repeat with 8chan, leading to Brennan's departure. See https://www.japantimes.co.jp/life/2014/03/20/digital/who-holds-the-deeds-to-gossip-bulletin-board-2channel

129. https://www.coindesk.com/markets/2019/11/06/93-days-dark-8chan-coder-explains-how-blockchain-saved-his-troll-forum and https://www.datacenterdynamics.com/en/analysis/uprooting-hate-8chans-founder-on-deplatforming-amazons-parler-ban-and-internet-extremism

131 The infrastructure of censorship in Asia

132 Gurshabad Grover   Gurshabad Grover is a technologist and legal researcher based in Bengaluru, India.

133 Keywords: Asia, internet censorship, politics, privatisation, surveillance

134   When people think about internet censorship in Asia, they likely think of the Great Firewall, which limits the sites Chinese users can visit and the content they can see. Like all archetypes, however, the Great Firewall is not representative of information controls everywhere. Asian governments seeking to restrict access to information they deem undesirable take a range of different approaches in how they implement this censorship infrastructurally. The Chinese government, for example, operates the Great Firewall directly, which means that the work of inspecting all traffic at international gateways is theirs 📄. Other Asian governments take a more

135   decentralised approach, tasking private actors with the work of censorship. The exact infrastructural implementation of these forms of government control enormously affects users' experiences of, and knowledge about, internet censorship.

Given that a more precise understanding will be key to more effective advocacy for a freely accessible internet, this piece will provide a broad tour of Asian infrastructures of censorship and highlight the features of decentralised information controls. If we begin with internet censorship in

Iran, often seen as the precocious cousin of the poster child that is the Great Firewall, we see that information controls are not *that* centralised. Iranian internet service providers (ISPs) have the choice to connect to international networks, bypassing infrastructures controlled by the state . These providers are still bound to follow state authorities' instructions when it comes to internet shutdowns and blocking of specific websites, but there is some diversity in infrastructural choices.

136

THE CASE STUDY OF INDIA
Or let's take India, where I live. Indian law requires that censorship orders be secret, which makes it difficult for Indian users to map the contours of repression 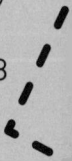. Not that knowing the government's rationale is always enlightening: TikTok, for instance, is deemed a threat to national integrity by the Indian government 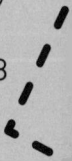. Unlike China, however, the Indian government has not made any effort to directly operate internet infrastructure for censorship. The government instead passes

137

138

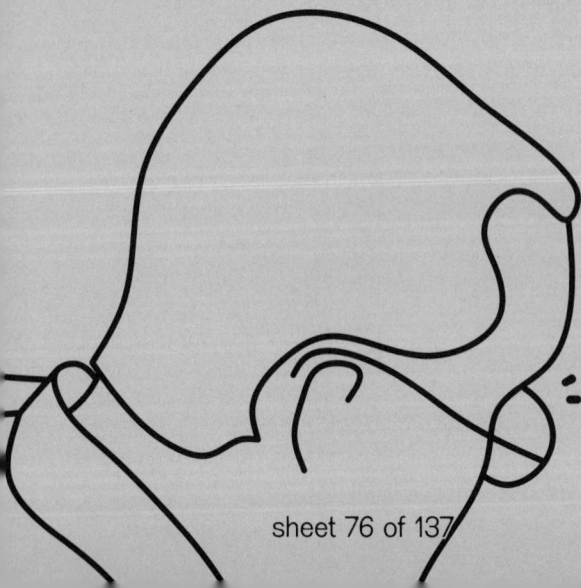

legal orders to ISPs. These information controls are 'decentralised' in the sense that it is the many public and private ISPs that are responsible for

139 → actually implementing the censorship 📠. There is no coordination mechanism amongst these ISPs, and they use whatever method their network engineers hacked up for the purpose. Different ISPs end up sometimes blocking different websites with different technical methods, making the experience of censorship across networks in India

140 → remarkably non-uniform 📠. Some are even using methods of censorship that make it technically

141 → impossible to present censorship notices 📠, leaving users confused and impeding challenges to unjust censorship.

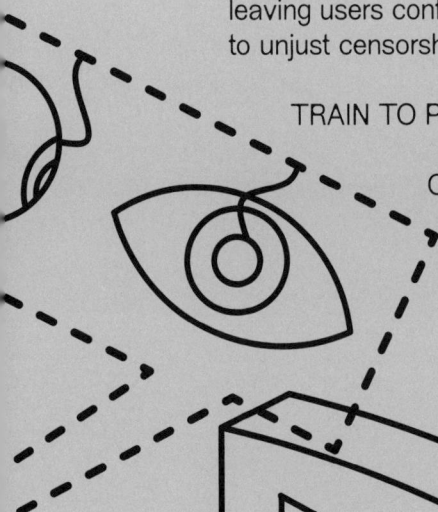

TRAIN TO PAKISTAN

Our neighbour, Pakistan, has a similar legal system to India's. Its implementation of censorship differs from India's,

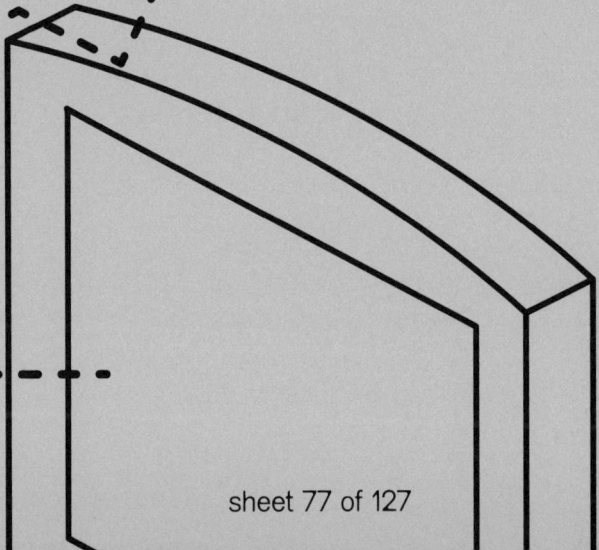

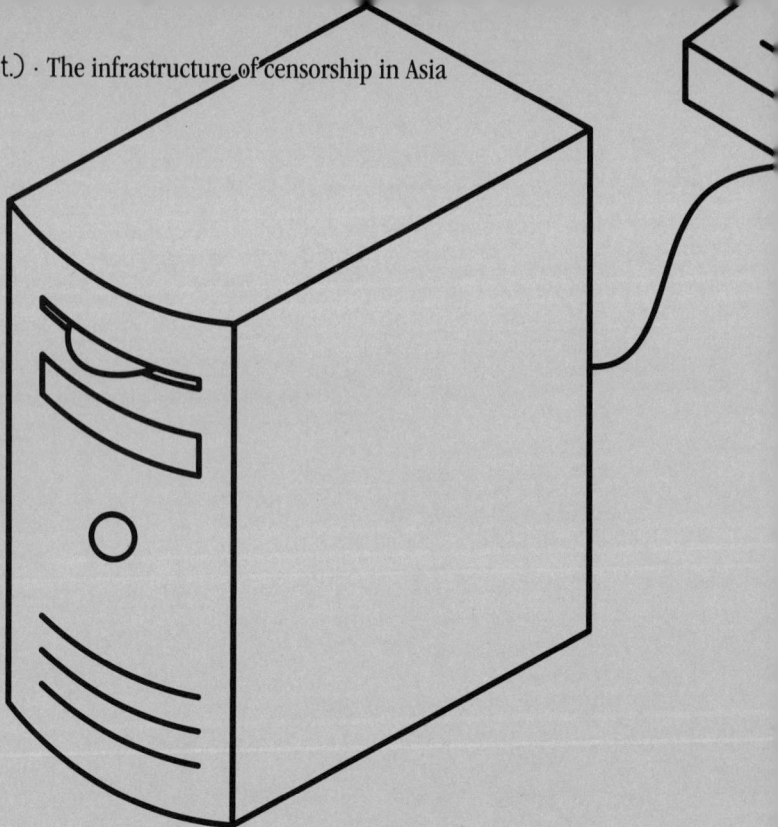

however, in that the Pakistan Telecommunications Authority (PTA) not only orders ISPs to block websites and webpages, but also deploys a centralised Web Monitoring System (WMS) to do the same itself 🖥. The WMS operates through state-controlled internet traffic exchanges, to which all ISPs are connected. All international connections flow through these exchanges, so this location provides a means for centralised control over traffic. To further centralise censorship mechanisms, in 2022 the PTA attempted to wrest control of all website lookups from ISPs 🖥. As it stands, information controls continue to exist both at the decentralised ISP level and at the level of centralised networks.

142

143

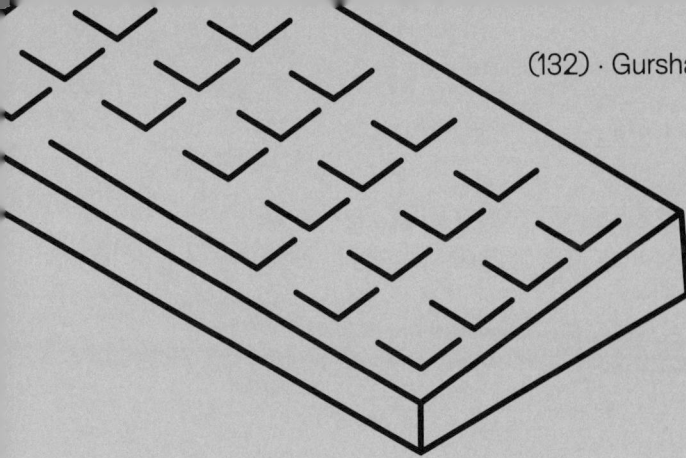

INDONESIA

Moving southeast, the network infrastructure in
Indonesia is a mix of private and public ISPs,
internet exchange points, and lines of international
connectivity 🖥. All the evidence suggests that
no centralised infrastructure exists or is used for
censorship. Instead, multiple state authorities can
send censorship orders to ISPs, which are then
responsible for implementing the blocking 🖥.
Unlike the Indian model, Indonesian censorship
includes technical guidance from the government
to ISPs in the form of regulations 🖥. The Ministry
of Communications and Information Technology
(MCIT) even maintains a central list of websites
that ISPs need to block. In theory, this would
result in decentralised blocking with central
coordination facilitated by the government. In
practice, however, the regulations allow for ISPs
to block 'negative content' based on their own
assessments 🖥. This results in ISPs blocking
websites and services beyond those specified by
the government, at their own discretion.

144

145

146

147

BREAK THE FIREWALL

If you want to see me on TikTok in the future, we will have to untangle the different technical, legal, and political roles of the states and infrastructure providers involved in censorship. So, what can we learn from these Asian cases and the differences between them?

First, internet surveillance and censorship mechanisms in a single jurisdiction can exist at multiple levels of infrastructural controls—at ISPs, internet exchange points, and lines of international connectivity. While information controls can be state-mandated, their implementation often falls on the shoulders of private companies. The decisions made by these private companies—from how they interpret orders to how they finally implement the blocking technically—exacerbate or minimise the effects of state-ordered censorship for the millions of users they serve.

Second, political context is key to understanding why governments decide to centralise or decentralise information controls. While government control over internet infrastructure may be loosening in Iran because of US sanctions that restrict the government's relationships with international networks 🗐 , Pakistan is trying to reduce its reliance on private companies and to centralise control in the state's hands. For Pakistan, this means increased costs in developing and maintaining infrastructure for information controls that will make surveillance and censorship immediate, unilateral, and uniform.

148 ⟵

Third, decentralisation is not always better for citizens' right to access information. It can just mean that more stakeholders have a say in the implementation of censorship, and private companies can choose to enforce censorship beyond what the government explicitly requires.

There is no one 'great firewall' in internet censorship. The better civil society, researchers, and activists understand the specific infrastructures involved, the better their advocacy for citizens' freedom will be.

---

149 Notes:

135. Deibert, R. 2010. China's Cyberspace Control Strategy: An Overview and Consideration of Issues for Canadian Policy. Canadian International Council. https://courses.cs.duke.edu/common/compsci092/papers/china/deibert2010.pdf

136. https://www.article19.org/wp-content/uploads/2020/09/TTN-report-2020.pdf

137. The Information Technology (Procedure and Safeguards for Blocking for Access of Information by Public) Rules, 2009. https://cis-india.org/internet-governance/resources/information-technology-procedure-and-safeguards-for-blocking-for-access-of-information-by-public-rules-2009

138. https://pib.gov.in/PressReleseDetailm.aspx?PRID=1635206#

139. Ramesh, R. et al. 2020. Decentralized Control: A Case Study of Russia. Network and Distributed Systems Security (NDSS) Symposium 2020. https://censoredplanet.org/assets/russia.pdf

140. Singh, K., Grover, G. and Bansal, V. 2020. How India Censors the Web. 12th ACM Conference on Web Science, 21-28. https://cis-india.org/internet-governance/how-india-censors-the-web-websci

141. Ibid.

142. https://www.pta.gov.pk/assets/media/tender_150218.pdf; https://www.codastory.com/authoritarian-tech/pakistan-web-monitoring-surveillance/

143. https://www.dawn.com/news/1693213

144. https://citizenlab.ca/2013/10/igf-2013-an-overview-of-indonesian-internet-infrastructure-and-governance/

145. https://freedomhouse.org/country/indonesia/freedom-net/2021

146. https://www.lexology.com/library/detail.aspx?g=99fc618d-232e-4590-b4cd-e456496e5ce9; https://www.lexology.com/library/detail.aspx?g=8fe08827-b0eb-4962-8f91-a6e2f7612515

147. https://ooni.org/post/indonesia-internet-censorship/

148. https://www.article19.org/wp-content/uploads/2020/09/TTN-report-2020.pdf

150 Encryption as a battleground in Ukraine

151 Ksenia Ermoshina — Ksenia Ermoshina is an Associate Research Professor at the Centre for Internet and Society of CNRS (the French national research council), in Paris, France.

152 Francesca Musiani — Francesca Musiani is an Associate Research Professor and Deputy Director at the Centre for Internet and Society of CNRS, in Paris, France.

153 Keywords: encryption, geopolitics, resistance, routing, Ukraine

154 To think about infrastructural politics today is to consider how internet infrastructures are defining contemporary geopolitical conflict. Russia's invasion of Ukraine in early 2022 instigated an intense battle for control over Ukrainian informational infrastructures on its temporarily occupied territories, including landlines, cables, and cell towers. The most intense period of this battle happened between May 30 and November 11, 2022, when Russian operator Miranda-Media took over the routing of internet traffic in territories controlled by Russia. This telecommunications company was initially established to aid Russian efforts in its occupation of Crimea, rerouting local traffic and making it easier for the Russian state to filter and censor the Ukrainian Internet. The very existence of this company, which eventually operated in other occupied parts of Ukraine, speaks to the key role of internet infrastructure in geopolitics.

Political use of internet infrastructure in war zones is more and more common. This makes encryption—an important but much less visible aspect of infrastructure than, for example, broken cables or shot down cell towers—an important battleground in Ukraine. By extension, this means digital security is one of the crucial components of the physical security of people living under Russian occupation. However, researching Ukrainian users' online practices in times of war shows that, paradoxically, advanced encryption is not always key to security. Rather, the war in Ukraine demonstrates that the physical and material functions of digital infrastructure, such as connectivity and internet access, can be more vital than privacy-enhancing technologies.

GENERAL DIGITAL SECURITY PRACTICES UNDER RUSSIAN OCCUPATION

All Ukrainians who remained in the country after the invasion are at risk of being disconnected, put under digital surveillance, and experiencing online censorship and filtering. Internet shutdowns, total or partial, are very frequent; infrastructures are targeted in bombing attacks because connectivity is a strategic resource in times of war. Being disconnected can mean the difference between life and death when digital communication tools are key in asking for food aid, medical help, electricity, and other key services. The conflict plays out through internet infrastructure. At the same time, some Ukrainian users face higher risks than others. We need to understand these differences through first-hand research. Stressing the nuances of connection and internet infrastructure politics under severe geopolitical stress helps us to provide adequate and regionally appropriate support.

Since the Russian invasion in February 2022, Ukrainians have lived in an 'asymmetric risk scenario', meaning that risk is not equally distributed across the population. Ukrainians living in occupied territories, for example, face problems like surveillance of their daily communications by Russian occupants, to which encryption seems an obvious solution. Using encrypted calls (over WhatsApp, for instance) to speak with relatives who live outside occupied territories, instead of relying on simple GSM calls, quickly became common practice for avoiding surveillance. Ukrainians under occupation also experience intense online censorship and disinformation campaigns pushing Russian propaganda.

In response, both specialised digital security trainers and ordinary citizens help spread Virtual Private Networks (VPN) and secure messaging applications, for example Signal, to protect the content of communications and circumvent censorship. Canadian NGO eQualit.ie, for example, set up servers in several major Ukrainian cities, including in places under Russian occupation. These servers allow people to communicate locally by using federated end-to-end encrypted tools, such as Element ▷ or Delta Chat ▷, which work even during internet shutdowns. As many Ukrainians in the occupied territories are involved in communicating sensitive information, like details about the position of Russian troops, considerable effort has been expended toward promoting these encrypted tools.

155, 156

Is encryption the answer for high-risk users?

Encryption is not always—or not only—the solution. Our research on the use of encrypted messaging systems in Ukraine, conducted between 2016 and 2019, focused on especially high-risk users, such as journalists, human rights defenders, and inhabitants of key battlefields like Crimea or Kherson ▷. These users' digital security practices 157 included secure messaging tools and other privacy-enhancing technologies. But rather than focusing on downloading encryption tools, such as Tor, or encrypted messaging services, such as Signal, high-risk users, and the digital security trainers that advise them, instead focus on developing tailored responses. Digital security trainers understand risk as highly contextual and rapidly changing, which makes digital security a multi-layered, social—rather than purely technical—process ▷.

158

This localised understanding of threat led us to conclude that more, or better, encryption is not always the answer for high-risk users, unlike what many advocates of privacy-enhancing technologies believe. In Kherson, an important port city under Russian occupation, Ukrainian users are at a high risk of device seizure. As a result, they consider relying on less commonly used or more technically sophisticated encrypted messengers a risk in its own right. The mere fact of having certain apps on one's phone (such as Signal, Tor or even Telegram) can raise suspicion and result in bodily harm or even life-threatening situations at the routine phone checks conducted by Russian soldiers.

In a strategy that might initially seem counterintuitive to those not in war situations, digital security trainers aware of this context advise their high-risk users to use WhatsApp and Gmail instead of Signal or a PGP-encrypted form of email. We also found that digital security trainers emphasise the importance of communicational autonomy and physical security, given that shutdowns and blackouts are the main concern of Ukrainian users both in occupied and unoccupied territories. Power banks, solar batteries, extra phones, mobile routers, and even Starlink antennas became the focus, alongside psychological self-help tutorials and first-aid classes. Encryption receded into the background, while the accessibility of communication became vital.

THE IMPORTANCE OF CONTEXT

For the specific threats that many high-risk users in occupied Ukraine face, the particularities of the Russian Occupation turn out to be key in defining people's communication practices. Or put more plainly, when it came to ensuring safe communication for those most at risk—like, for example, the occupied peoples of Kherson—the familiarity with and inconspicuousness of particular tools proves to be a more crucial advantage than the privacy protections guaranteed through better encryption technologies.

The Ukrainian approach to security underlines that risk is relational and local. Security should be considered a multi-layered complex process, in which the digital layer is just one of many. The practices of Ukrainian users teach us that the protective potential of encryption is always and intrinsically linked to physical, psychological, and operational politics—as well as infrastructural concerns. Our thinking of infrastructural politics—or security in war situations—should always begin with this fact.

---

159 Notes:

155. https://element.io
156. https://delta.chat/fr
157. Ermoshina, K. and Musiani, F. 2022. *Concealing for Freedom: The Making of Encryption, Secure Messaging and Digital Liberties.* Manchester, UK: Mattering Press.

158. See Ermoshina and Musiani, op. cit.; see also Kazansky, B. 2021. 'It depends on your threat model': the anticipatory dimensions of resistance to data-driven surveillance. *Big Data & Society* 8 (1). https://doi.org/10.1177/2053951720985557

Encryption regulation, and what to do about it?

161 Mallory Knodel

Mallory Knodel is the Chief Technology Officer for the Center for Democracy and Technology in Washington DC and is a member of the Internet Architecture Board. Mallory is a human rights advocate with a focus on encryption, censorship, and cybersecurity.

162 Keywords: anti-encryption, chilling effects, civil society, global regulation, risk

163 Encryption is a promising feature of digital communication infrastructures. In a world increasingly reliant on digital connection, encryption creates a protective boundary between outsiders and those who are exchanging information online. It guarantees confidentiality, integrity, and authenticity of information exchange. Encryption technologies are essential to commerce, national security, and

164 civic life. End-to-end encryption ⚷ —from here on just 'encryption'—is also key to protecting human rights. It has an obvious function in protecting privacy. Yet other rights are also secured, both online and offline, by strong encryption: freedom of opinion and expression; the right to access information; and free association and the right to protest—to name a few.

The use and implementation of encryption for communication infrastructures—like the Signal app or encrypted email or internet access through Virtual Private Networks (VPNs)—is increasingly under siege. Governments across the world are undertaking anti-encryption efforts, leaning on age-old claims of 'lawful intercept' and 'exceptional access'. This trend is worrying as restricting encryption means widening the possibilities for government surveillance and control. Analysing global anti-encryption policy trends and offering concrete strategies to chart out the path forward for human rights defenders in the digital age is key to answering the question posed by this chapter's title.

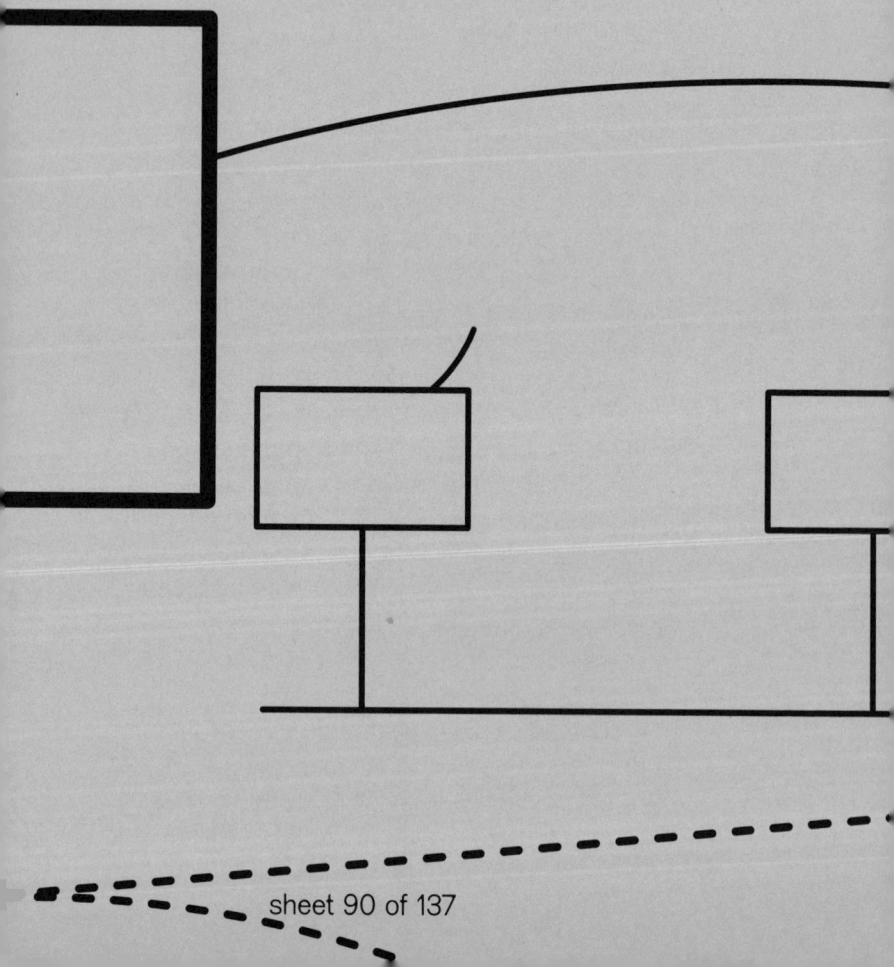

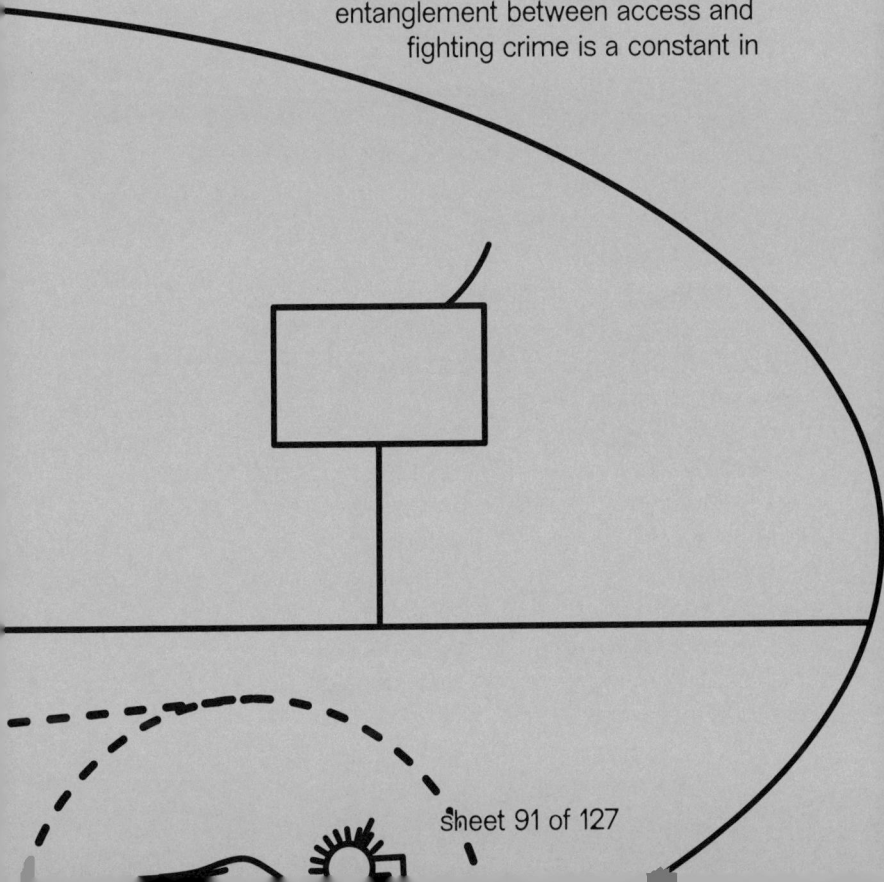

ENCRYPTION AT RISK

Breaking encryption is a site of international cooperation. Law enforcement and the intelligence community are driving legislative efforts that have aimed to bring encryption to heel, in authoritarian regimes and democracies alike, for decades. A renewed effort became apparent when, in 2018, the 'Five-Eyes' intelligence alliance, which comprises Australia, Canada, New Zealand, the United Kingdom, and the United States, joined forces with India and Japan in calling on corporations to intentionally weaken encryption so that law enforcement agencies might gain 165 ⟶ exceptional access to communications ⟨⟩. This entanglement between access and fighting crime is a constant in anti-encryption policy proposals across the globe, although the justifications for and the exact scope of access vary.

The following six telling examples help elucidate the spectrum of attacks on encryption and how they undermine rights and civil liberties:

Australia: Perhaps the first, and still singular, example of a mandated encryption 'backdoor' is Australia's Telecommunications and Other Legislation Amendment of 2018 ⚷. It directly exempts the government from data protection responsibilities and expands its authority by requiring Australian companies to ensure lawful access to data, irrespective of the use of encryption for user confidentiality and privacy.

166

Zimbabwe: During protests in 2014, the government sowed distrust and confusion by claiming that it could access messages on the then-newly encrypted platform WhatsApp. Ultimately the government simply blocked all access to the popular platform, a disproportionate action that exacerbated social instability ⚷. In this context, destabilising trust in encryption stifled speech and dissent.

167

Ethiopia: In 2014, the government jailed bloggers for participating in a workshop on the use of end-to-end encryption because it was considered a 'clandestine' activity ⚷. In this framing, encryption becomes a form of crime or potential crime that warrants forceful government interventions—demonstrating the real-life consequences of encryption legislation.

168

India: In 2020, the government required that encrypted platforms be 'traceable'. The euphemism of traceability mandates that encrypted platforms keep track of metadata: who has sent and received messages, for example. While this avoids direct targeting of content, and at first sight does not seem to break encryption, traceability techniques directly weaken the entire platform and proliferate user data.

Europe: The Council of Europe proposed in 2020 that encrypted systems 'open up' in service to the fight against child sexual abuse materials online. The proposed regulation, and ones that followed it more recently, specify that all digital service providers are responsible for preventing the sharing of illegal content, but not how they should do so. This requirement stands even if the provider cannot see the content, as is the case for end-to-end encrypted systems, and led proponents to suggest the use of client-side scanning, a dangerous development where content would be scanned on a device before it is encrypted.

169   United States: The US has been a central site of the decades-long Crypto Wars ⚇. These debates have come to a head after Apple refused to decrypt the phone of a mass shooter in 2015. In response to this refusal, the US Government presented a policy proposal that would allow law enforcement to compel decryption and device

170   scanning when accessing encrypted systems ⚇.

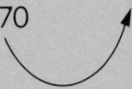

LEGISLATING MAGICAL THINKING

Anti-encryption proposals overlook the fact that by targeting the architectural design of encrypted environments, any single exception weakens the security and privacy of all users, everywhere, all the time. Europe's proposal to limit encryption to hold digital service providers responsible for the spread of criminal content, for example, codifies a widespread surveillance regime ⚷. Moreover, the technical means to comply with traceability and exceptional access alike does more than just reveal a message's content: it amplifies the need for commercial surveillance. Modern-day 'backdoors', implemented at scale, require the tracking and storage of far more metadata than platforms currently collect on their users. This means that anti-encryption legislation always ends up giving companies increased responsibility to manage, analyse, and access user data—at odds with end-user privacy gains.

171

'Magical thinking' is happening when the object of legislation is both the problem—encryption is protecting criminals—and at the same time it is touted as the solution—encryption backdoors will help law enforcement fight crime. It is therefore not difficult to see that legislation banning end-to-end encryption is inconsistent with human rights principles of necessity and proportionality of law enforcement interventions. From the blunt approaches in Ethiopia and Zimbabwe to the tech-savvy proposals in the EU and US, to the softer requirement of metadata tracking in India: they ultimately have a similarly chilling effect.

MITIGATING THE RISKS

Encryption opponents benefit from the technical complexity of encryption and sow uncertainty in whether traceability, client-side scanning or other proposed techniques break it. They do, in fact, under a strict and succinct definition of strong encryption. Civil liberties advocates should remain confidently unwavering in support of users' expectations of confidentiality and privacy when they are using encrypted applications. Major civil liberties issues are at stake in the anti-encryption debate; with client-side scanning and compelled decryption putting the police in our pockets without oversight. Civil society must confront, rather than avoid, the tension between privacy and safety inherent in law enforcement in the digital age. End-to-end encryption is a key component of the internet's functioning—and as such, and important topic for civil society to monitor in global debates about technology policy.

---

172 Notes:

164. End-to-end encryption is a strict application: the 'ends' are defined, usually as sender and receiver, and everyone else is kept out of the exchange.

165. https://www.justice.gov/opa/pr/international-statement-end-end-encryption-and-public-safety

166. https://www.internetsociety.org/news/press-releases/2021/new-study-finds-australias-tola-law-poses-long-term-risks-to-australian-economy

167. https://www.hrw.org/news/2016/07/07/dispatches-zimbabwe-blocks-internet-amid-police-crackdown

168. https://advox.globalvoices.org/campaigns-research/behind-bars-in-ethiopia-campaign-to-free-the-zone9-bloggers

169. https://en.wikipedia.org/wiki/Crypto_Wars

170. https://www.law.georgetown.edu/american-criminal-law-review/wp-content/uploads/sites/15/2020/03/57-1-compelled-decryption-and-state-constitutional-protection-against-self-incrimination.pdf

171. https://cdt.org/insights/briefing-document-on-key-issues-in-european-commissions-csam-proposal

PART 3

CONSUMED BY

THE
COMPUTER?

175 critical infrastructure walks in Amsterdam

176 Niels ten Oever

Niels ten Oever, is an Assistant Professor of AI and European Democracies at the European Studies Department and co-founder of the critical infrastructure lab at the University of Amsterdam.

177 Maxigas

Maxigas is a Senior Lecturer in New Media and Digital Culture at the Media Studies Department and co-founder of the critical infrastructure lab at the University of Amsterdam.

178 Keywords: antennas, cell towers, cityscapes, sensors, walking

179 We have all said it at least once: 'The internet isn't working'. But do we know what happens when a web page fails to load, or our messages don't go through? Through what? What makes the internet work? What devices are the culprit? This chapter is an invitation to join us, the researchers of the critical infrastructure lab at the University of Amsterdam, on a critical infrastructure walk to explore communication and control infrastructures in the city. Infrastructure tends to 'fade into
180 the woodwork of society' 📡 but careful and persistent observation and interrogation can render them visible, and lead to hard questions about accountability and the structuration of public space.

Walking helps us see differences across the city: open areas with few antennas located on cell towers, or dense urban landscapes with antennas on electricity masts at every crossing (so that they capture signals in all directions!). Being attentive to the 'radioscape' around us opens the city in new ways: who gets coverage, where do networks become dense and where do they become spotty, to what extent do devices like antennas and cell phone towers yield to or resist the natural or built environment, and what areas of the city are dense in infrastructure for police and government surveillance? Or in other words: if we know how to look, the city's digital infrastructure, and the powers that shape it, stand out.

WALKING AS METHOD

Walking as a critical practice to understand how the internet is used for quotidian control and surveillance in the city is not new. Alison Powell uses the data walking method 📡 to understand how climate data is collected in London 📡. Ingrid Burrington wrote a practical field guide that helps New Yorkers inhabit their city through attention to the networks they traverse 📡. Karin van Es and Michiel de Lange rely on walks to understand the datafication of cities more broadly, to contextualise and make data infrastructures visible 📡. Walking gave each of these thinkers a way to perceive their world anew, and in this sense, this unorthodox method can set the political imagination abuzz.

181
182
183
184

Walking empowers participants to understand the logics of infrastructural and architectural control in cities, and critically question them. Our walks in Amsterdam look at key sensor and communications technologies, like routers and antennas, as well as infrastructures of surveillance, including CCTV

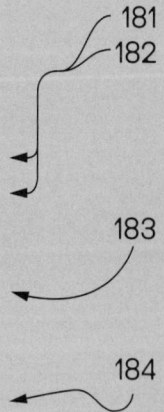

cameras. During the walks, participants use their eyes, ears, binoculars, spectrum analysers, databases, and smartphones to see how people and data flows are organised within public spaces—meaning that our method can be applied to a wider range of networks and cities.

WAKING UP TO INFRASTRUCTURE

What would it mean to 'wake up' to the infrastructures your familiar devices, like your phone, connect to regularly? How do we turn the tables on the surveillance potential of these networks and devices, as we connect to them through our personal devices? Walking in digital infrastructures opens densely layered global networks that are both vital to our everyday lives and hard to conceptualise. On an infrastructure walk, we begin with networking devices: tangible, concrete, and essential to the vibrations of air that traverse the globe. As the critical practice of walking makes so clear: we must begin with what we can see.

WALKING UP TO INFRASTRUCTURE

When we take people on a tour of Amsterdam, we help them 'unfade' the infrastructure from the woodwork: a city dotted with antennas, cameras, sensors, actuators, and other key instruments that make up the city's communication and control networks. We work with several tools that help us see differently. Most of the tools we use are widely available and cheap to access, like the Cell Mapper or the Architecture of Radio apps, and even the tinySA spectrum analyser—in case you are

interested in developing your own infrastructure
walks. These tools help
expand the range of
what we can

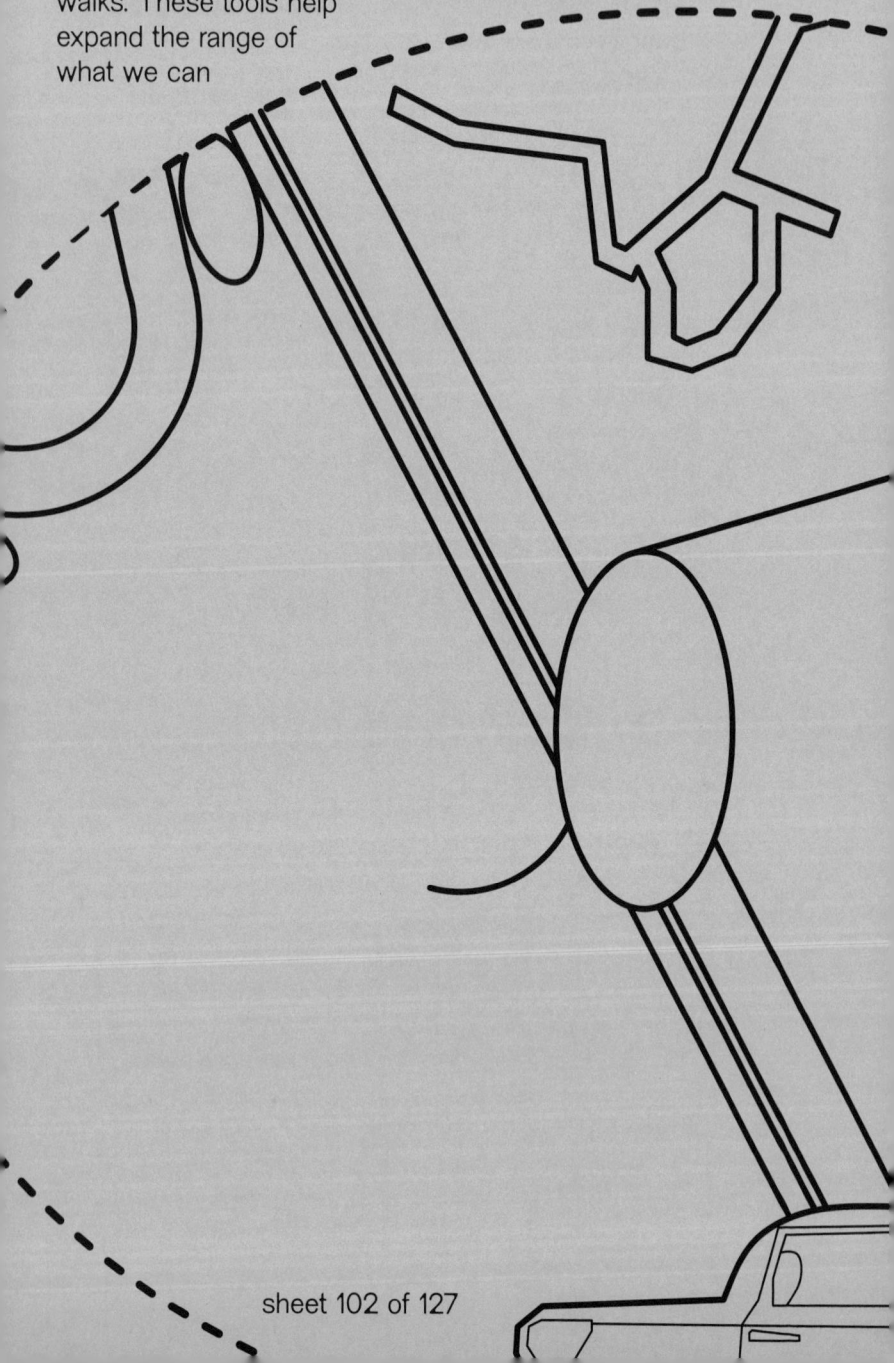

see, and our understanding about, for example, the location and ownership of the antennas that carry our data. The biggest part of what we do, however, is change people's mindsets, by directing them to see the city anew through an infrastructural lens. This takes not much more than a different way of looking, sometimes augmented by binoculars.

Infrastructure shapes our action possibilities, but we rarely know who the power players in the field are. Sometimes all the tools we need to understand who these actors are is our eyes. During the walks, we examine the labels of the antennas that collect and send out the city's data. These can indicate who owns them or give away the make and model of the device. If you know the model, you know the frequency at which the antenna operates, and this tells you about the purpose of the device: is it intended for military, police, meteorological, or civilian use?—and its function 📶. If it is part of a small local network, you can see where the data is being sent: a police station, for example. Knowing what to look for

185

creates a vantage point from which to think anew about the intersections between convenience and surveillance in the city.

NEXT STEPS

In line with Burrington, seeing our surroundings differently is full of political potential. A walk might help you imagine the data streams that flow over devices you see and the electromagnetic waves you can sense, with some technical support. Only once these technologies stand out to you, suddenly stark, can you begin to imagine how these flows might run differently. A different infrastructure would create different flows; different priorities and different needs would send signals in different directions. Who are these infrastructures serving? Should there be places where there is no data flowing at all? These are important political questions about the infrastructures that are part of our everyday environment, but you cannot begin to answer them until you start walking.

---

186 Notes:

180. Bowker, G.C. and Star, S.L. 2000. *Sorting Things Out: Classification and Its Consequences.* Revised edition. Cambridge, MA: The MIT Press. See also https://www.ics.uci.edu/~gbowker/classification

181. http://lifewinning.com/projects/networks-of-new-york

182. van Es, K. and de Lange, M. 2020. Data with Its Boots on the Ground: Datawalking as Research Method. *European Journal of Communication* 35 (3) 278–89. https://doi.org/10.1177/0267323120922087

183. See for example: for the US, https://external-preview.redd.it/kHz8LypfoS ReUFHUPODXYyxR022CllqWkToFMaquW gA.jpg?auto=webp&s=bd164a2bec898eaf8 10856594db93e48472115fd or this one for the Netherlands: https://globaldigitalcultures. org/wp-content/uploads/2022/04/Maxigas-Picture-3.png

184. http://lifewinning.com/projects/networks-of-new-york

185. See note 3.

The human in the machine

---

188 trust in internet governance

189 Ashwin Mathew

Ashwin Mathew is a Lecturer in Global Digital Cultures in the Department of Digital Humanities at King's College London.

---

190 Keywords: internet governance, risk, routing, social ties, trust

---

191 The internet could hardly exist without trust. From 'trust and safety' teams in large social media companies to trustworthy computing, trust has become a weasel word: we assume that we all mean the same thing when we use it, but 'trust' becomes increasingly polysemous, covering senses that are variously regulatory, technological, institutional, and interpersonal. Yet, even as there is substantial focus on the importance of trust to the use of the internet, attention to the critical role that trust plays in the governance of internet infrastructure remains largely unexamined.

Internet infrastructure is founded upon trust, with interpersonal trust playing an essential function. This stands in stark contrast to other global infrastructures (such as airlines or postal systems), which are governed primarily through the interaction of market interests with national and international state regulation. It also stands in contrast to trust-free technological models—such as those promulgated by blockchain advocates—based on the claim that if only we can get the technology right, we will increase individual freedom by doing away with the need to trust.

But while trust can be hidden away, it can never be made to go away—to create trustworthy technological systems, it's essential to understand how, where, and why trust builds, between whom, and how it is integrated in internet governance. Understanding this, allows us to see the human in the internet's machinery.

RISKY BUSINESS: TRUST AND GOVERNANCE

To analyse the role of trust in internet infrastructure, we must first arrive at definitions of trust and related terms to avoid the confusion of different senses of trust. We can gain more clarity on what we mean by trust by re-framing these issues in terms of the antecedent to trust: risk. Rather than asking whether a technology, person, or institution can be trusted, we might instead ask 'how can the risks of relying on this technology, person, or institution be mitigated?' Living with risk and striving for its minimisation involves people and social relationships just as much as it does technologies or institutions.

We generally deal with risk through structures of *assurance*, ranging from legal regimes to cultural norms that shape how people behave in risky situations ⬭. In economic exchange, for example, ⟵ 192 there is always a risk that money will lose its value. This is mitigated through the assurances provided by central banks. However, risks can never be completely mitigated by assurance structures. A central bank might assure the value of money, but the choice to engage in economic exchange always entails risks it cannot cover, such as the behaviour of partners. Contracts and legal regimes might act as assurances about the behaviour

of partners,
but these are
typically invoked only
when a relationship
fails. Assurance
structures are essential
to mitigating risk in the
complexity of modern
193   life ◯, but we must be
attentive to their limits.

This is where *trust* becomes
relevant. Trust allows us to take
the leap of faith necessary to engage in
194   social interactions that have the potential to go
awry ◯. Trust is the social phenomenon that
enables cooperation despite risk: it is what allows
humans to continue working together when no
further assurances (legal or otherwise) are available.
To trust is to make a conscious choice to engage
in social interaction. *Interpersonal trust* emerges in
dyadic social relations, in which people choose to
trust one another ◯. Without conscious choice,
195   it could be said that there is *confidence* in the
196   outcome of a risky interaction, rather than trust ◯.
This is the situation with many assurance
structures, where switching costs are sufficiently
high as to make any choice unreasonable. Returning
to the example of the central bank, few people can
choose the monetary system in which they interact.

It could be said that they have confidence in the system, but not trust. The limits of assurances, and the sociality of trust, raise the question: how is trust built in internet infrastructure?

TRUST IN INFRASTRUCTURAL PRACTICE

There is no real choice for ordinary internet users but to have confidence in assurance structures. For example, multistakeholder governance ◯ is 197 often held to be among the key innovations of internet governance, in bringing assurance through the participation of an array of stakeholders, contested though this might be. Similarly, even while cryptocurrencies claim to be trust-free—with no need to trust a central bank—they rely upon assurances provided by cryptographic algorithms and decentralised networks of transaction validators. However, as important as assurance structures are for internet governance, confusing assurance with trust ◯ blinds us to the inherently human nature of 198 governance through interpersonal trust.

The internet owes its ongoing existence to the work of technical communities of network administrators and cybersecurity engineers who operate and maintain internet infrastructure across the world. In these transnational professional communities, obligations to peers—held in interpersonal trust relations—are just as important as economic obligations held as employees of companies, or political obligations held as citizens of states.

Cybersecurity engineers, for example, routinely face the dilemma of how to responsibly share information about newly discovered vulnerabilities and attacks. On the one hand, this information should be shared openly to ensure widespread response; on the other hand, sharing such information raises the significant risk that it might be exploited by malicious actors. In the global environment of the internet, cybersecurity threats demand a collaborative response, as they cannot be contained within corporate or state boundaries. Several government and industry bodies play critical roles in inter-organisational and cross-national cybersecurity cooperation, acting as assurance structures for sharing sensitive information. However, mindful of the responsibilities that they hold, cybersecurity engineers engage in cooperation with colleagues who they trust individually, even while they also rely on these assurance structures. Indeed, the greater the risk, the more likely that cybersecurity engineers will cooperate and share sensitive information with individuals (rather than organisations) who

199 → they trust to be responsible, and who can aid in collaborative response ⬭.

In the interconnection of the tens of thousands of individual networks that comprise the global internet, network administrators constantly take risks in assuming responsible behaviour among peers who manage the other ends of network interconnections. The risks in managing network interconnections cut across corporate and national boundaries, making it difficult to mitigate them through assurance structures, such as contractual agreements or regulatory arrangements. Instead, it is interpersonal trust that enables ongoing, rapid coordination among network administrators to mitigate risk, ensuring the reliable operation of the global internet ⬭.

200

201

Interpersonal trust is integral to the operation of the global internet ⬭. Clearly understanding how this trust works reveals new and unusual political possibilities. We have a robust understanding of assurance structures in internet governance; what would it mean to build on and elaborate on the role of interpersonal trust in governing internet infrastructure?

---

202 Notes:

192. Yamagishi, T. and Yamagishi, M. 1994. Trust and Commitment in the United States and Japan. *Motivation and Emotion* 18 (2) 129–166.

193. Giddens, A. 1990. *The Consequences of Modernity*. Stanford University Press.

194. Luhmann, N. 1979. *Trust and Power*. John Wiley and Sons.

195. Hardin, R. 2002. *Trust and Trustworthiness*. Russell Sage Foundation Publications.

196. Luhmann, N. 1988. Familiarity, Confidence, Trust: Problems and Alternatives. In Gambetta, D. (ed.) *Trust: Making and Breaking Cooperative Relations*. Basil Blackwell, pp. 94–107.

197. Raymond, M. and DeNardis, L. 2015. Multistakeholderism: Anatomy of an Inchoate Global Institution. *International Theory* 7 (3) 572–616.

198. Cheshire, C. 2011. Online Trust, Trustworthiness, or Assurance? *Daedalus* 140 (4) 49–58.

199. Mathew, A.J. and Cheshire, C. 2018. A Fragmented Whole: Cooperation and Learning in the Practice of Information Security. Technical Report. UC Berkeley Center for Long-Term Cybersecurity and Packet Clearing House. https://www.pch.net/resources/Papers/A_Fragmented_Whole/

200. Mathew, A.J. 2014. *Where in the World is the Internet? Locating Political Power in Internet Infrastructure*. University of California, Berkeley. http://www.ischool.berkeley.edu/research/publications/ashwin_mathew/2014/where_world_internet_locating_political_power_internet_infrastructure

201. Also see: Mathew, A.J. and Cheshire, C. 2017. Risky Business: Social Trust and Community in the Practice of Cybersecurity for Internet Infrastructure. In Proceedings of the 50th Hawaii International Conference on System Sciences. Hawaii International Conference on System Sciences, Waikoloa, Hawaii, USA, pp. 2341–2350. https://doi.org/10.24251/HICSS.2017.283; Mathew, A.J. 2022. Can Security Be Decentralised? In Parkin, S. and Viganò, L. (eds) *Socio-Technical Aspects in Security*. Cham: Springer International Publishing (Lecture Notes in Computer Science), pp. 67–85. https://doi.org/10.1007/978-3-031-10183-0_4

---

204 the state of civil society's participation at the ITU

205 Mehwish Ansari

Mehwish Ansari is Head of Digital at ARTICLE 19, an international human rights organisation that works to protect and promote freedom of expression and access to information.

206 Keywords: China, civil society, geopolitics, internet standards, ITU

---

207   Digital technology standards are having a moment in international politics. In 2021, the G7—an informal group of countries representing some of the largest economies in the world, including the US, EU, and Japan—announced a framework for collaboration on developing standards for the internet, telecommunications, and emerging technologies that support 'democratic values', 'freedom', and 'open societies'. In 2023, the Office of the United Nations (UN) High Commissioner for Human Rights presented a report at the request of the UN Human Rights Council on how to protect and promote human rights in technical standards development for new and emerging digital technologies.

Ultimately, both initiatives identified that to effectively uphold these values and rights, governments need to address the lack of representation of civil society in standards development, particularly in intergovernmental organisations like the International Telecommunication Union (ITU). But fixing this problem requires more political will and action from governments than they are currently showing.

THE STATUS QUO OF ITU
STANDARDS DEVELOPMENT

A standard is a written technical documentation that sets out specifications, requirements, operations, or terminology that guide technology design, development, or deployment. The technologies we interact with every day—from the Wi-Fi networks in our homes to the video surveillance cameras in our public spaces—are fundamentally shaped by the norms, assumptions, and prescriptions set in standards. This is why standards development matters: the chance to influence a technical standard is the chance to change a technology before it becomes embedded in the real world.

The ITU works on international matters of telecommunication, information, and communications networks. One of the agency's most significant responsibilities is setting standards for these technologies. As a UN specialised agency, the ITU's standards development is largely driven by governments, or Member States. In recent years, the ITU has invested heavily in growing its industry membership to improve the technical legitimacy of its standards. In contrast, the ITU does not officially recognise civil society organisations as stakeholders in its standards development.

When questioned at an open forum on technical standard-setting and human rights in 2021, the Chief of Study Groups in the ITU's Technical

Standardisation Sector claimed that there are no barriers to civil society organisations because they can always opt to become Sector Members, the membership designation for industry. While this sounds inclusive, sector membership is not actually structured to encourage civil society participation. The high fees for Sector Members are designed for private companies and therefore prohibitive for non-profit organisations, and there is no clear option for requesting fee waivers. Moreover, an organisation applying for sector membership can be rejected by the Member State where the organisation is based, without any option for appeal or review. This leaves civil society organisations at the mercy of the same governments they seek to hold accountable.

Without a feasible path to membership, the extremely few civil society organisations that participate in ITU standards development do so as part of government delegations, through Member States that have made commitments to civil society inclusion, including through the G7 and Human Rights Council. However, under these terms, civil society may become caught in broader geopolitical dynamics that threaten its independence and effectiveness.

CIVIL SOCIETY: A PAWN IN
A GEOPOLITICAL GAME?

The recent interest of the G7 in standards development is a response to China's increasing dominance in global technology markets. As part of China's 'Made in China 2025' strategy, Chinese government institutions and industry have invested heavily in standardising their technologies, and they have deliberately focused on the ITU because of its strong reputation in the Global South. In conversations with ITU participants, representatives of governments in Asia and Africa, such as Malaysia and Egypt, have explained that they often feel alienated or overlooked in other standards organisations that are dominated by US or European industry interests. They respect ITU standards because they believe the ITU respects them.

The US, EU, and their allies often frame their competition with China in the technology sector as an ideological battle between democracy and authoritarianism, and their calls for free, open, and human rights-respecting technical standards in the ITU is part of this rhetoric. Of course, these governments hold some genuine commitment to these values. Since 2022, the US and EU have introduced millions of dollars in new funding opportunities for civil society to advocate for human rights in the ITU and other standards organisations.

However, these governments are also undoubtedly motivated by their own economic interests. While the US and UK block proposals by Chinese industry to standardise facial recognition in the ITU by citing fundamental human rights concerns, they continue to deploy facial recognition

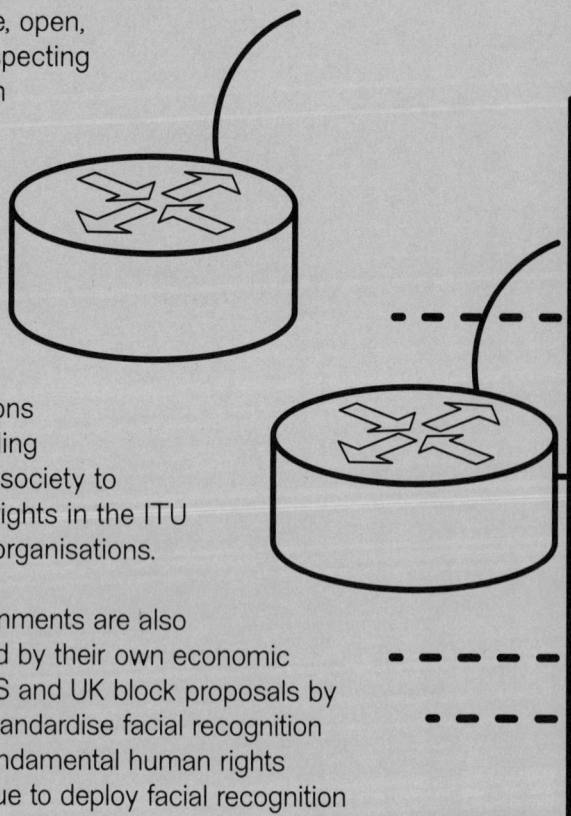

technologies made by US or European companies in domestic policing and other highly sensitive contexts. Unless they are consistently and universally applied, human rights and other public interest values are nothing more than a shield to protect North American and European industry from competition.

These dynamics raise fundamental questions about civil society's role and effectiveness. What happens when civil society advocates' priorities in the ITU don't align with efforts to counter China— or even come into conflict with the economic interests of the governments they depend on to participate? Ultimately, they must support all the positions of the Member State delegations they join, as a condition of their participation—even if the positions are inconsistent with human rights.

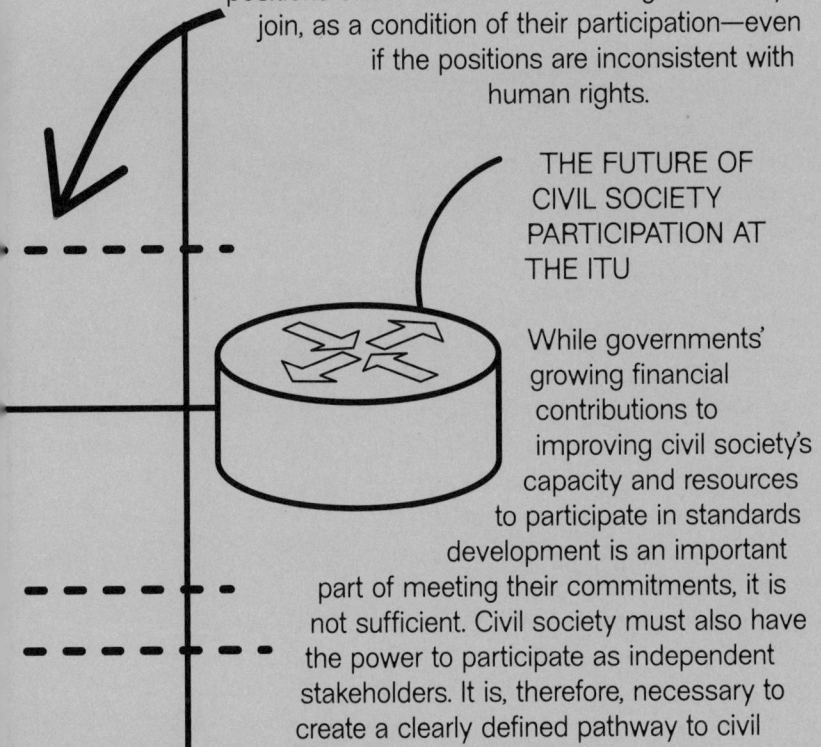

THE FUTURE OF CIVIL SOCIETY PARTICIPATION AT THE ITU

While governments' growing financial contributions to improving civil society's capacity and resources to participate in standards development is an important part of meeting their commitments, it is not sufficient. Civil society must also have the power to participate as independent stakeholders. It is, therefore, necessary to create a clearly defined pathway to civil society membership at the ITU that does not impose fees or other financial obligations and ensures that the review process for membership applications is subject to transparency and accountability. However, this change can only happen with the agreement of the agency's most powerful decision-makers: its Member States.

The US, UK, EU, and other members of the G7 and Human Rights Council must bring and defend concrete proposals to improve pathways to ITU membership for civil society. Held every four years, the ITU's Plenipotentiary Conference functions as its supreme governing body and is the most important opportunity for making this change. However, if states are truly committed to supporting civil society, they must not wait. In 2022, a new ITU Council was elected from among the ITU's Member States, including many that have committed to improving civil society inclusion. As the ITU's governing body between Plenipotentiary Conferences, the Council is a critical forum for raising, discussing, and implementing changes that improve civil society inclusion.

It is clear that formally recognising civil society as an ITU stakeholder will be an uphill battle, especially against Member States that have been responsible for the growing efforts to shrink civic space and threats to civil society. However, difficulty is not a reason to forgo effort. The standard practice of the ITU needs to change. If the members of the G7 and Human Rights Council are serious about supporting civil society in spaces like the ITU, they must take responsibility for enacting this change.

209 ## learning from social movements

210 Jenna Ruddock — Jenna Ruddock is a Research Fellow with the Technology and Social Change Project at the Harvard Kennedy School's Shorenstein Center.

211 Joan Donovan — Joan Donovan is the Research Director of the Shorenstein Center on Media, Politics and Public Policy at the Harvard Kennedy School's Shorenstein Center.

212 Keywords: Cloudflare, networked social movements, privatisation, public internet

213
214 What do companies care about, and how can we know? Internet infrastructure is dominated by a small cohort of large private corporations. Their decision-making rarely makes headlines, but when it does, these companies often disavow the politics of infrastructure and the power they consequently exert, while preaching the gospel of neutrality. In 2019, cloud computing and security company Cloudflare came under fire for its provision of infrastructural services to 8chan, a largely unmoderated and deeply hateful message board implicated in several mass shootings. In response to calls for Cloudflare to stop doing business with the site, CEO Matthew Prince insisted that—unlike social media platforms—infrastructure services are 'mere conduits' for online content, whose core mission is
215 'to help build a better internet'.

What Prince failed to clarify was: a better internet for who? For the public? For the company's shareholders? For paying customers? For advertisers? Where do those visions overlap— and, more tellingly, where do they diverge? We propose a new way to conceptualise and describe how the internet could better serve the public interest, drawing from social movement technology practices, where rules and resources are organised to ensure equity in critical infrastructure.

INFRASTRUCTURES ARE NOT NEUTRAL

Infrastructure has always been more than a 'mere conduit'. A site of power and contestation, infrastructure can never be an apolitical domain  —and internet infrastructure isn't any less political than roads, water utilities or railways . But internet infrastructure is notably opaque. The internet writ large is sold as an ethereal, nonmaterial technology: from 'cyberspace' to the 'cloud', it exists both 'everywhere and nowhere' . This framing is itself a political act—it obscures everything from the physical presence of internet infrastructure in our communities and the labour necessary to operate and maintain it, to decisions about costs and extractive demands on vital resources . The politics of internet infrastructure are often further submerged beneath concerns about freedom of expression and fears of misguided regulation.

But what the public can't see, we can't scrutinise or subject to public oversight—or collectively reimagine. Even as internet infrastructure companies continue to publicly deny their

216

217

218

219

political power, other critical infrastructures are increasingly becoming internet-reliant—not only our communications networks, but also our healthcare systems, energy grids, water treatment facilities, and key supply chains. We need an honest discussion about what drives these infrastructural politics.

DISTINGUISHING BETWEEN 'FOR PROFIT' AND 'FOR THE PUBLIC GOOD'

The 8chan case, as a rare occasion of public decision-making by an infrastructure provider, offers a glimpse of the politics at play in the day-to-day provision of infrastructural services. Despite dramatic declarations of reluctance, Cloudflare ultimately did remove 8chan from its networks and would do the same to other websites in other cases, all while continuing to insist that its role should only be to act as one of the conduits that a 'better internet' will require. Like other infrastructure companies before and since, Cloudflare acted how it did and, importantly, *when* it did because its decisions are *not* solely motivated by the depoliticised vision of a 'better internet'.

Because of its organisational form—corporate, private, for-profit—its decision-making must prioritise the interests of investors, shareholders, and paying customers, which may or may not align with any vision of 'a better internet' for the rest of us. It's exactly because of these obligations that infrastructure companies can never—and will never—act as 'mere conduits' of content online. Cloudflare similarly cited its dominant market share to bolster its claim that it should remain neutral in cases of harmful actors utilising its services . These kinds of claims evade any discussion of how Cloudflare—and others like it—profits from bad actors, and further fails to reflect on whether that dominance and scale is itself good for the internet-at-large.

220

WHAT CAN WE LEARN FROM NETWORKED SOCIAL MOVEMENTS?

We already have the tools to conceptualise an internet that serves the broadest public good. The internet needs to be grounded in an overt set of politics that places people over profit and aims for equitable distribution of services and products, where inequality and sustainability are taken as seriously as technological bugs. From local and regional collectives to global networked movements, communities are already paving the way to make these politics real, building public interest tools and developing alternative governance models that are practitioner-led, self-hosted, and non-commercial.

What can we learn from how these networked movements decentralise their leadership and the tools they use? Critically, their approach to infrastructure is explicitly political , prefacing community maintenance and ownership that is subject to collective oversight, unlike that of commercial actors. Furthermore, these movements leverage existing communication infrastructure to spread out their messages and resources, so that no one corporation has a restrictive hold on them. They host, build, and contract their own infrastructure, or rely on third parties that have a not-for-profit purpose, like Signal, while prioritising knowledge- and capacity-building, from creating

221

222 educational zines to hosting teach-ins . Or in other words, social movements are providing clear models for building a 'better internet'.

A PUBLIC INTEREST INTERNET

Infrastructure is politics by other means. Systemically disrupting the dominance of private, for-profit internet infrastructure requires a more widespread and urgent willingness to confront the politics of internet infrastructure. We see this in how infrastructure corporations invest resources, expand services, and approach target markets and regions (or don't). These run from straightforward corporate interests—like a desire to avoid bad press—to more nuanced organisational biases, such as a willingness to cooperate with law enforcement (or not), absent legal requirements to do so. Furthermore, without strong regulation, technology corporations are free to give or restrict services and access according to the beliefs of their CEOs—a vanishingly small cohort given the concentration of corporate power within internet infrastructure. This reflects the dangers of letting corporate internal political logics shape the governance of critical infrastructure.

Corporate capture; a lack of public education, outreach, and accountability; innovators raising capital to foster the development of technologies while the impacted public and policy trails behind—these are all well-known themes in the history of communication technologies. Yet today is different: the internet is more than a collection of communication technology products. It is key infrastructure revolutionising societies much beyond communications, rapidly. We know that what companies care about won't deliver us an internet that answers to the public. Alternative governance models and infrastructures are possible—they exist right now. We see networked social movements developing infrastructures that are rooted in the public's interest. We must apply their lessons to provide concrete alternatives to corporate 'conduits'.

223 Notes:

214. Moriniere, S. 2023. From cyber to physical space: the concentration of digital and data power by tech companies. Open Data Institute. https://www.theodi.org/article/from-cyber-to-physical-space-the-concentration-of-digital-and-data-power-by-tech-companies

215. https://blog.cloudflare.com/terminating-service-for-8chan

216. Star, S. and Ruhleder, K. 2021. Steps Toward an Ecology of Infrastructure. *Informations Systems Research* 4.

217. Ensmenger, N. 2021. The Cloud is a Factory. Your Computer Is on Fire 29.

218. https://www.eff.org/cyberspace-independence

219. See for example Tarnoff, B. 2022. *Internet for the People: The Fight for Our Digital Future* 38; Philip, K. 2021. The Internet Will Be Decolonized. *Your Computer is on Fire* 97; Ensmenger, N. 2018. *The environmental history of computing*, Regalado, A. 2011. Who coined 'Cloud Computing'? *MIT Tech. Rev.* https://www.technologyreview.com/2011/10/31/257406/who-coined-cloud-computing

220. https://blog.cloudflare.com/terminating-service-for-8chan

221. See for example https://urbanomnibus.net/2019/10/building-the-peoples-internet

222. See for example https://instituto.mijente.net/courses/tech-wars

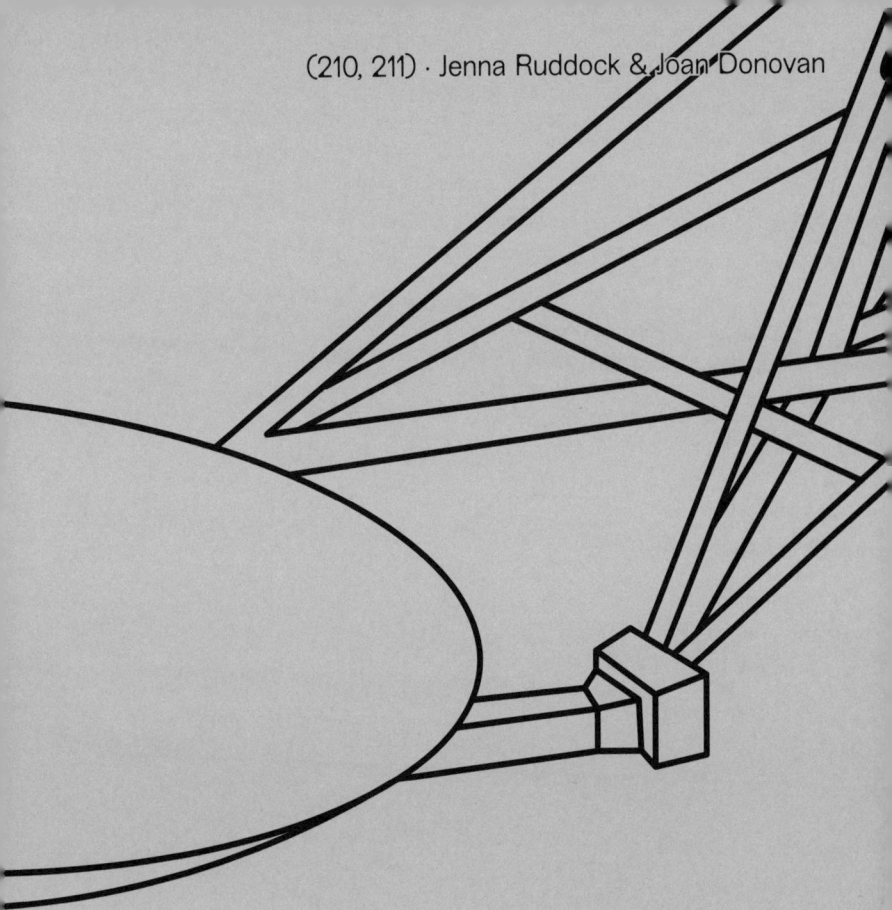

225

Perhaps I bit off more than I could chew, as covering a topic as vast as the politics of internet infrastructure is a tall order. As such, I have many people to thank: this book would not have come into being without the online and 'meat space' support from Mark Graham and Joe Shaw at Meatspace Press. I also want to thank David Sutcliffe and Suzanne van Geuns for their support in editing this book, as well as John Philip Sage and Carlos Romo-Melgar for the beautiful design. I am very grateful for the funding I received from the Ford Foundation (no. 136179, 2020) which helped make this book a reality, from pixel to print. I am especially indebted to Ford's Michael Brennan, Lori McGlinchey, and Jenny Toomey. Last but not least, I want to thank all the contributors for their wonderful chapters and concrete suggestions for how to ensure internet technologies enable, rather than eat, us.

The design of *Eaten by the Internet* explores the
patent document as a graphic medium where the
experience of technological infrastructure begins.
Patents govern possibilities, define their boundaries
and nominate their owners.

227

Through documenting intellectual property,
patents become instruments to frame knowledge
and orchestrate its reproduction, exploitation,
expansion or modification. As noted by Gitelman,
these 'epistemic objects' occasionally exist for
'their potential to show' in the future, rather than
merely as archives of knowledge. • Patents impose
a vague outline of what could transpire upon their
'enactment'. It is this vagueness, dressed up as
accuracy, that makes these documents a rich
graphic medium for visual experimentation.

228

The knowledge of the genre is not limited to its
'literary' component, but also on the ways that it
is given a format: arrows, numbered references,
systematic content hierarchies, suggestive
abstracted illustrations. In Lee's words, referring
to documents, 'their pedestrian visual language
often falls at the periphery of what the market and
professional designers tend to think of as "graphic
design"', making them immutable by means of
their 'banality'. •

229

Exploiting this (questionable) form of legitimisation,
the book deploys a patent-like referencing
framework where arrows are ubiquitously
present, everything is sequentially numbered
and referenceable. The layout of *Eaten by the
Internet* offers a neurotic performance of accuracy,

parodying the different graphic devices that constitute the patent as a visual form of knowledge production. To punctuate this relentless document, the illustrations adopt the method of kitbashing: the illustrations of relevant patents are exploded into parts, 'disorienting' • them from their original intent, offering readers an opportunity to establish alternative associations with the text.

230

231 Notes:

228. Gitelman, L. 2014. *Paper Knowledge: Toward a Media History of Documents.* Duke University Press Books.
229. Lee, C. 2022. *Immutable: Designing History.* Onomatopee.
230. Ahmed, S. 2006. *Queer Phenomenology: Orientations, Objects, Others.* Duke University Press Books.

Other publications by Meatspace Press include:

Fake AI
Data Justice and COVID-19: Global Perspectives
How to Run a City Like Amazon, and Other Fables
Towards a Fairer Gig Economy
Our Digital Rights to the City

All Meatspace Press publications listed above are free to download,
or can be ordered in print from meatspacepress.com